UMAP
ILAP

Modules
2002–03

Tools for Teaching

published by

The Consortium for Mathematics
and Its Applications, Inc.
Suite 210
57 Bedford St.
Lexington, MA 02420

edited by

Paul J. Campbell
Campus Box 194
Beloit College
700 College St.
Beloit, WI 53511–5595
campbell@beloit.edu

ISBN 0–912843–74–8

Typeset and printed in the U.S.A., using LaTeX on an Apple Macintosh computer.

Table of Contents

COMAP

Introduction

The instructional Modules in this volume were developed in two projects undertaken by COMAP and sponsored by the National Science Foundation: the Undergraduate Mathematics and Its Applications (UMAP) Project and the Interdisciplinary Lively Applications Project (ILAP).

UMAP

Project UMAP develops and disseminates instructional modules and expository monographs in mathematical modeling and applications of the mathematical sciences, for undergraduate students and their instructors.

UMAP Modules are self-contained (except for stated prerequisites) lesson-length instructional units. From them, undergraduate students learn professional applications of the mathematical sciences. UMAP Modules feature different levels of mathematics, as well as various fields of application, including biostatistics, economics, government, earth science, computer science, and psychology. The Modules are written and reviewed by instructors in colleges and high schools throughout the United States and abroad, as well as by professionals in applied fields.

UMAP was originally funded by grants from the National Science Foundation to the Education Development Center, Inc. (1976–1983) and to the Consortium for Mathematics and Its Applications (COMAP) (1983–1985). In order to capture the momentum and success beyond the period of federal funding, we established COMAP as a nonprofit educational organization. COMAP is committed to the improvement of mathematics education, to the continuing development and dissemination of instructional materials, and to fostering and enlarging the network of people involved in the development and use of materials. In addition to involvement at the college level through UMAP, COMAP is engaged in science and mathematics education in elementary and secondary schools, teacher training, continuing education, and industrial and government training programs.

ILAP

ILAPs are small interdisciplinary group projects designed to motivate mathematical concepts and skills. They differ from UMAP Modules, which teach content through exposition, examples, and exercises, in being *problem-solving* projects.

An ILAP contains

- background information and introduction to the situation,

- project handouts,

- requirements to be fulfilled in a solution, and

- other supporting materials (sample solutions, further background material, notes for the instructor) as appropriate.

Like case studies, these projects often require students to use scientific and quantitative reasoning, mathematical modeling, symbolic manipulation, and computation to solve problems, analyze scenarios, understand issues, and answer questions. The level of prerequisite skills varies; the final product is an oral presentation or a written report.

Other Materials

In addition to this annual collection of UMAP Modules, other college-level materials distributed by COMAP include individual Modules (more than 500), *The UMAP Journal*, and UMAP expository monographs. Thousands of instructors and students have shared their reactions to the use of these instructional materials in the classroom, and comments and suggestions for changes are incorporated as part of the development and improvement of materials.

Acknowledgments and Vade Mecum

This collection of Modules represents the spirit and ability of scores of volunteer authors, reviewers, and field-testers (both instructors and students). The substance and momentum of the UMAP Project comes from the thousands of individuals involved in the development and use of UMAP instructional materials. COMAP is very interested in receiving information on the use of Modules in various settings. We invite you to call or write for a catalog of available materials, and to contact us with your ideas and reactions. A Guide for Authors is at the end of this volume.

Sol Garfunkel, COMAP Director
Paul J. Campbell, Editor

UMAP

Modules in Undergraduate Mathematics and Its Applications

Published in cooperation with

The Society for Industrial and Applied Mathematics,

The Mathematical Association of America,

The National Council of Teachers of Mathematics,

The American Mathematical Association of Two-Year Colleges,

The Institute for Operations Research and the Management Sciences, and

The American Statistical Association.

Module 783

Game Theory Models of Animal Behavior

Kevin Mitchell
James Ryan

Applications of Biology

COMAP, Inc., Suite 210, 57 Bedford Street, Lexington, MA 02420 (781) 862–7878

INTERMODULAR DESCRIPTION SHEET: UMAP Unit 783

TITLE: Game Theory Models of Animal Behavior

AUTHOR: Kevin Mitchell
Department of Mathematics and Computer Science
Hobart and William Smith Colleges
Geneva, NY 14456
mitchell@hws.edu

James Ryan
Department of Biology
Hobart and William Smith Colleges
Geneva, NY 14456
ryan@hws.edu

MATHEMATICAL FIELD: Game theory

APPLICATION FIELD: Biology

TARGET AUDIENCE: Students in either a game theory course or an introductory course on animal behavior.

ABSTRACT: This unit is an introduction to elementary game theory and some of its applications to evolutionary biology. The concept of an evolutionary stable strategy (ESS) is defined and its consequences are explored in several two- and three-person games. References are made throughout to examples of contests between animals in the wild. The unit concludes with a detailed application of this theory to male elephants and their mating strategies, using data from research studies.

PREREQUISITES: None.

UMAP/ILAP Modules 2002–03: Tools for Teaching, 1–48.
©Copyright 2003 by COMAP, Inc. All rights reserved.

COMAP, Inc., Suite 210, 57 Bedford Street, Lexington, MA 02420
(800) 77-COMAP = (800) 772-6627, or (781) 862-7878; http://www.comap.com

Game Theory Models of Animal Behavior

Kevin Mitchell
Department of Mathematics and Computer Science
Hobart and William Smith Colleges
Geneva, NY 14456
mitchell@hws.edu

James Ryan
Department of Biology
Hobart and William Smith Colleges
Geneva, NY 14456
ryan@hws.edu

Table of Contents

MODULES AND MONOGRAPHS IN UNDERGRADUATE
MATHEMATICS AND ITS APPLICATIONS (UMAP) PROJECT

The goal of UMAP is to develop, through a community of users and developers, a system of instructional modules in undergraduate mathematics and its applications, to be used to supplement existing courses and from which complete courses may eventually be built.

The Project was guided by a National Advisory Board of mathematicians, scientists, and educators. UMAP was funded by a grant from the National Science Foundation and now is supported by the Consortium for Mathematics and Its Applications (COMAP), Inc., a nonprofit corporation engaged in research and development in mathematics education.

Paul J. Campbell   Editor
Solomon Garfunkel   Executive Director, COMAP

1.   Introduction

1.1   Animals Playing Games

As cooler temperatures descend on the Rocky Mountains during the fall, male elk (*Cervus elpahus*) enter a state of heightened sexual activity referred to as "rut." A rutting male spends considerable time calling to attract females into a harem that the male defends from other males. When another male attempts to mate with a female from the harem, the harem holder races in to drive off the interloper. In some cases, these disputes lead to fighting between the two rival males, and the loser may be wounded during the contest. Even the winner may fare poorly in the long run, as the constant battles to protect his harem often leave him weak and in poor condition at the onset of winter [McCullough 1969].

Animal conflicts also occur over access to food resources. A group of vultures feeding on a dead wildebeest, for example, squabble with one another over access to the carcass. In this case, the resource being contested is food and not access to females, but the rules of the contest are essentially the same. At first, it might seem that we could apply optimality models to animal conflicts by determining the costs and benefits and solving for the optimal strategy. Unfortunately, this is not possible, because the optimal strategy for one individual depends on the behavior of its competitors in the population.

Evolutionary game theory is adapted from economic theory, except that the currency is Darwinian fitness and not money [von Neumann and Morgenstern 1944; Maynard Smith and Price 1973; Maynard Smith 1982]. Game theory is similar to optimization theory (e.g., the study of optimal foraging) in that it attempts to identify the best strategy based on the costs and benefits of some required resource. There are, however, important differences. Unlike optimization theory, where an individual's reproductive success depends only on its own behavior, game theory involves more than one contestant and a contestant's success depends on the behavior of all the other players.

1.2   Rules of the Game

In the simplest terms, economic game theory deals with two or more players each attempting to select the best response to the anticipated strategy of their opponents. Evolutionary game theory, however, tends not to focus on individual players, but on strategies available to different categories of individuals (i.e., male vs. females, dominants vs. subordinates, experienced vs. inexperienced, etc.). Each type of player has a set of strategies that it can adopt in response to its opponent's anticipated strategy. Players that adopt the best strategies contribute more offspring to future generations and therefore by definition have higher reproductive fitness. If the best strategy is also heritable, it will become the dominant strategy for this category of player over evolutionary time. In theory, the game stabilizes when all categories of players have adopted their best response to each of their opponents.

As Mesterton-Gibbons and Adams [1998] point out, in evolutionary game theory the "best" solution to the contest is the strategy that can be expected to evolve by natural selection. In other words, any possible alternative behavior would yield lower reproductive fitness or else it would have already spread throughout the population. This "best" solution is termed an *evolutionary stable strategy* (ESS). Informally, an ESS is a strategy that cannot be replaced by any (rare) alternative strategy that appears in the population (i.e., a mutant strategy). An ESS may be pure (consist of a single strategy) or mixed (consist of several strategies in a stable equilibrium).

The fitness of a genotype (sometimes) depends on the genetic composition of the population (it is "frequency-dependent"). In the language of game theory, this means that the payoff from a particular strategy depends on the strategies of the other players, so multiplayer game models are needed. In most cases, evolutionary games are played within a single species by a local population in a specific environment. In addition, game theory models allow an individual to play more than one strategy over time. Multiple strategies allow animals to modify their behavior in different conditions. A harem-holding male elk, for example, would behave differently if it were displaced and had to take on the role of intruder.

1.3   Types of Evolutionary Games

A game is a model describing the behavioral interactions of two or more individuals whose interests may conflict. In order to model the potentially conflicting interactions between players, we must be precise about

- who is involved in the game,

- what "moves" are possible, and

- how one player's success depends on the behavior (moves) of the other players in the game [Hammerstein 1998].

In other words, a game has

- a set of players or categories of players;

- a strategy set—a list of alternative behaviors or morphologies that each category of player could use; and

- a set of payoffs (in terms of evolutionary fitness) for each possible combination of strategies.

Virtually all evolutionary games can be classified based upon four criteria. Different combinations of these four criteria yield different types of games each with its own set of assumptions, methods, and ESSs [Bradbury and Vehrencamp 1998]. The four criteria are:

- type of strategy set,

- type of player symmetry,

- number of opponents at one time,

- the number of sequential decisions in the game.

A strategy set can include discrete strategies (e.g., fight or flee), continuous strategies (e.g., vary the frequency of a call over a range of possible frequencies), or some combination of behaviors (e.g., produce pheromone A 30% of the time and pheromone B 70% of the time). If the strategy contains only one behavior, it is called a *pure strategy*. When a player (or category of players) performs a combination of alternative strategies, the combination is called a *mixed strategy*. For example, male frogs often call to attract mates. Suppose there are two genotypes for calls and 75% of male frogs always give call A (one genotype) while 25% percent of males always give call B (second genotype). In a second species of frogs, each male can perform each call type and each male uses call A 75% of the time and call B the remaining 25% of the time. Both are examples of a mixed strategy (in each case, females encounter a 3:1 ratio of calls).

Games can also be classified by the type of player symmetry. If the game is symmetrical, then all players use identical strategy sets and players are essentially interchangeable. Two male warblers of identical age and size displaying over access to a suitable nest site would be playing a symmetrical game if each could use the same set of display behaviors. Asymmetrical games, however, are probably more common in animal populations. Asymmetrical games have at least two categories of players, with each category having "access to different alternative strategies, different probabilities of winning with a given strategy, different payoffs when they win with a given strategy, or some combination of these conditions" [Bradbury and Vehrencamp 1998]. Interactions between dominant and subordinate individuals within a social group, or between males and females, are asymmetrical, because the players are not likely to use the same strategy set and there are different fitness payoffs to each type of contestant.

A simple *contest* involves a player and a single opponent at a time, such as two male yellow warblers disputing the boundaries of a territory. In contrast, *n-person games* or *scrambles* involve more than two players at once. Imagine a harem-holding bull elk defending his harem against a series of bachelor males. Here the possible payoffs to the bull depend on the frequency with which alternative strategies are used by the bachelor males. If all bachelors elect to fight, the payoffs are different than if only 40% elect to fight and the other 60% elect a strategy of sneaking copulations when the harem master is otherwise occupied. In other words, payoffs for *n*-person games are frequency-dependent.

A discrete symmetric contest requires a different model from an asymmetric *n*-person game. For each type of game, there are unique assumptions, different sets of ESSs possible, and different ways of finding those ESSs. In this article. we discuss contests involving two players.

2. Two-Strategy Games

We denote the *expected payoff* to a contestant playing strategy A against another playing strategy B by $E(A, B)$. Note that in general $E(A, B) \neq E(B, A)$, since the payoffs to different strategies are likely to be different. Ordinarily this information is presented in a matrix (see **Table 1**) that gives the fitness payoffs for each cell in the matrix.

Table 1.
The payoff matrix for Player 1 in a two-strategy game.

Player 1	Player 2	
	A	**B**
A	$E(A, A)$	$E(A, B)$
B	$E(B, A)$	$E(B, B)$

The "best" course of action depends on what the other players are doing. A gull, for example, might feed itself by two strategies: Catch its own fish or steal a fish from another gull. If almost all gulls forage by catching their own fish, then it pays to be a thief as there are many potential victims. If most of the gulls are playing the thief strategy, then it may not pay to steal food.

2.1 Discrete Symmetric Contests:
The Game of Chicken

Most readers are familiar with the game of Chicken, at least from the movies. In our version, two players bet a certain amount of money and then drive their cars towards each other. There are two strategies. A Chicken will swerve at the last second to avoid an accident, but in the process will lose the bet. A player who is Not Chicken will not swerve out of the way and will crash into the other player if that player also does not swerve.

Clearly the payoff to each player (strategy) depends on the opponent's strategy. To calculate the payoff in a contest, we need to know:

- the value of the prize or resource,

- the cost of winning,

- the probability of winning,

- the cost of losing, and

- the probability of losing.

The payoff to strategy A when playing against strategy B is given by

$$E(A, B) = \text{(probability that } A \text{ beats } B) \times \text{(value of resource} - \text{cost of winning)}$$
$$- \text{(probability that } A \text{ loses to } B) \times \text{(cost of losing)} \qquad \textbf{(1)}$$

Suppose that each player bets \$100 on the race and that it takes \$10 to get the car ready for the race. Suppose further that it costs \$1000 to repair a car in the event of a crash. Let C denote the Chicken strategy and N the Not Chicken strategy.

When two Chickens race against each other, we assume that there is a 50% chance that either contestant swerves first and thus loses. The value of the contest is \$200, the amount of money in the pool. It costs each contestant \$100 that they bet and \$10 to prepare the car, for a total of \$110 whether the contestant wins or loses. Thus, using **(1)**, we have

$$E(C, C) = 0.5 \times (200 - 110) - 0.5 \times (110) = -10.$$

This makes sense, since C wins half the time and it always costs \$10 to prepare the car. In a Chicken vs. Not Chicken contest, the Chicken always loses, so

$$E(C, N) = 0 \times (200 - 110) - 1 \times 110 = -110.$$

In a Not Chicken vs. Chicken contest, Not Chicken always wins, so its payoff is

$$E(N, C) = 1 \times (200 - 110) - 0 \times 110 = 90.$$

When two Not Chickens play each other, neither swerves and an accident results. We'll assume that each gets their \$100 back, but it still cost \$10 to prepare the car and \$1,000 to repair the car. Therefore,

$$E(N, N) = 0 \times (200 - 110) - 1 \times (10 + 1000) = -1010.$$

It is convenient to gather these payoffs into a single matrix, as in **Table 2**, which shows the payoffs for Player 1. Notice that $E(N, C) \neq E(C, N)$: The first quantity is the payoff to an N when playing a C and the second is the payoff to a C when playing an N; there is no reason that they should be the same.

Table 2.

The payoff matrix for Player 1 in the game of Chicken.

Player 1	Player 2	
	Chicken	Not Chicken
Chicken	−10	−110
Not Chicken	90	−1010

In this game, there are two *pure strategies*, namely, always play Chicken (swerve) or always play Not Chicken (never swerve). But some contestants might play a *mixed strategy*, that is, a mixture of the Chicken and the Not Chicken strategies. The choice of which strategy to play might depend on peer pressure, the reputation of the opponent, whether or not the car belongs to one's parents, and so on. While **Table 2** gives the payoffs for pure strategies only, with a bit of algebra it can be used to calculate the payoffs for mixed strategies.

Suppose that A represents a mixed strategy that plays Chicken with probability p and Not Chicken with probability $1 - p$. We denote this by the linear combination

$$A = pC + (1 - p)N.$$

Similarly, let B be a second mixed strategy such that

$$B = qC + (1 - q)N.$$

If we assume that A and B choose their strategies independently in each encounter, then the probabilities of the various strategic encounters are multiplicative. There will be a Chicken vs. Chicken encounter with probability pq, a Chicken vs. Not Chicken encounter with probability $p(1 - q)$, a Not Chicken vs. Chicken encounter with probability $(1 - p)q$, and a Not Chicken vs. Not Chicken encounter with probability $(1 - p)(1 - q)$. Each of these probabilities must be multiplied by the expected payoff for the corresponding encounter and then added to give the total payoff $E(A, B)$. Thus, to evaluate $E(A, B)$, we simply expand the expression as if it were a product of factors:

$$\begin{aligned} E(A, B) &= E(pC + (1 - p)N, qC + (1 - q)N) \\ &= pqE(C, C) + p(1 - q)E(C, N) + (1 - p)qE(N, C) + \\ &\quad + (1 - p)(1 - q)E(N, N). \end{aligned} \tag{2}$$

For example, if A plays Chicken 60% of the time (so $p = 0.6$) and B plays Chicken 20% of the time (so $q = 0.2$), then using **(2)** and the values in **Table 2**, we have

$$\begin{aligned} E(A, B) &= (0.6)(0.2)(-10) + (0.6)(0.8)(-110) + (0.4)(0.2)(90) \\ &\quad + (0.4)(0.8)(-1010) = -370. \end{aligned}$$

Which strategy is best? The answer depends on the population of opponents. If a town consisted of almost all Chickens, then a Not Chicken would win most contests. On the other hand, if a town consisted of almost all Not Chickens, the Chicken strategy would be more prudent; while it would not win many contests, it would never result in an accident costing $1,010. Intuitively, we might expect that there is some mix of Chicken and Not Chicken such that the average payoff to each strategy is precisely the same. That is, there should be some stable mixture of strategies so that any player would be disadvantaged by switching from one strategy to the other.

Definition 1 *A strategy S is an evolutionary stable strategy (ESS) if for every strategy*
T ≠ S we have

$$E(S, S) \geq E(T, S)$$

and if $E(S, S) = E(T, S)$, *then*

$$E(S, T) > E(T, T).$$

This definition was originally given by Maynard Smith [1974]. For S to be an ESS, the first condition says that no strategy has a higher payoff against S than S itself. When there is a strategy T that has the same payoff as S does against S, then the second condition says that S has an advantage over T when playing T. In evolutionary terms, the strategy S can invade any population, since it will out-compete any other strategy T. Moreover, no strategy T can invade a population of S, since S will out-compete it. In this sense, a population of S-strategists is stable. Chicken is not an ESS, because

$$E(C, C) = -10 < 90 = E(N, C).$$

But neither is Not Chicken an ESS, because

$$E(N, N) = -1010 < -110 = E(C, N).$$

Is there some mixture of the Chicken and Not Chicken strategies that is stable? How to find such mixtures is the subject of the next section.

Exercises

1. a) What is the value of $E(B, A)$ in the payoff matrix in **Table 3**?

Table 3.
The payoff matrix for **Exercise 1**.

Player 1	Player 2	
	A	B
A	0	2
B	-3	4

b) Is either strategy an ESS?

2. Suppose in a game of Chicken that it costs x dollars to get the car ready for the race, that each player bets y dollars, and that it costs z dollars to repair the car in case of accident. Write out the general payoff matrix for this game.

2.2   ESSs and Two-Strategy Games

We now consider a general two-strategy game. We show that such a game always has an ESS. Denote the two strategies by X and Y. **Table 4** shows the payoffs for this general situation. We assume that the strategies are distinct, that is, we do not have both $a = c$ and $b = d$.

It is easy to check for pure ESSs: X is pure ESS if either $E(X, X) > E(Y, X)$ or if both $E(X, X) = E(Y, X)$ and $E(X, Y) > E(Y, Y)$. Using the payoff matrix, this is equivalent to either $a > c$ or ($a = c$ and $b > d$). Similarly, Y is an ESS if either $d > b$ or ($d = b$ and $c > a$). Consequently, there is no pure ESS only when both $c > a$ and $b > d$.

Now suppose that there is no pure ESS. We will (eventually) show that there must be a mixed strategy

$$S = pX + (1 - p)Y$$

that is an ESS, where p denotes the probability with which strategy X is played and $1 - p$ the probability with which strategy Y is played. We first examine the properties that any such mixed ESS must have.

Theorem 1 *Let $S = pX + (1 - p)Y$ be a mixed ESS, where X and Y are pure strategies. Then the payoff to X or to Y against S is the same as the payoff to S against itself. That is,*

$$E(X, S) = E(Y, S) = E(S, S).$$

Moreover, if $T = qX + (1 - q)Y$ is any mix of the X and Y strategies, then

$$E(T, S) = E(S, S).$$

Proof. First, we have

$$E(X, S) = pE(X, X) + (1 - p)E(X, Y), \quad E(Y, S) = pE(Y, X) + (1 - p)E(Y, Y).$$

From **(2)**, we get

$$
\begin{aligned}
E(S, S) &= p^2 E(X, X) + p(1 - p)E(X, Y) + (1 - p)pE(Y, X) + (1 - p)^2 E(Y, Y) \\
&= p[pE(X, X) + (1 - p)E(X, Y)] + (1 - p)[pE(Y, X) + (1 - p)E(Y, Y)] \\
&= pE(X, S) + (1 - p)E(Y, S). \tag{3}
\end{aligned}
$$

Table 4.

The payoff matrix for a general two-person game.

Player 1	Player 2	
	X	Y
X	a	b
Y	c	d

But S is an ESS, so $E(S,S) \geq E(X,S)$ and $E(S,S) \geq E(Y,S)$, so **(3)** becomes

$$E(S,S) = pE(X,S) + (1-p)E(Y,S) \leq pE(S,S) + (1-p)E(S,S) = E(S,S).$$

Since the first and last terms are equal, the middle relation must also be an equality. Consequently, $E(S,S) = E(X,S)$ and $E(S,S) = E(Y,S)$. Using these equalities, we get

$$
\begin{aligned}
E(T,S) &= qpE(X,X) + q(1-p)E(X,Y) + (1-q)pE(Y,X) \\
&\quad + (1-q)(1-p)E(Y,Y) \\
&= q[pE(X,X) + (1-p)E(X,Y)] + (1-q)[pE(Y,X) + (1-p)E(Y,Y)]. \\
&= qE(X,S) + (1-q)E(Y,S) \\
&= qE(S,S) + (1-q)E(S,S) \\
&= E(S,S).
\end{aligned}
\tag{4}
$$

$\square$

Corollary 1 *Let X and Y be any pure strategies and let $S = pX + (1-p)Y$ be any mixed strategy such that $E(X,S) = E(Y,S)$. Then*

$$E(X,S) = E(Y,S) = E(S,S).$$

Proof. As in **(3)** in the proof of **Theorem 1**,

$$E(S,S) = pE(X,S) + (1-p)E(Y,S).$$

But since $E(X,S) = E(Y,S) = E(S,S)$, then

$$E(S,S) = pE(X,S) + (1-p)E(X,S) = E(X,S) = E(Y,S). \qquad \square$$

By **Theorem 1**, if $S = pX + (1-p)Y$ is a mixed ESS, then $E(X,S) = E(Y,S)$, that is, $pE(X,X) + (1-p)E(X,Y) = pE(Y,X) + (1-p)E(Y,Y)$. Using the payoffs in **Table 4**, this means

$$pa + (1-p)b = pc + (1-p)d.$$

Collecting the terms involving p yields

$$b - d = p(b+c-a-d), \tag{5}$$

so that

$$p = \frac{b-d}{b+c-a-d}. \tag{6}$$

Since we have assumed that there is no pure ESS, $c > a$ and $b > d$. So $0 < p < 1$, in other words, p is a legitimate probability.

Conversely, if there is no pure ESS and $p = (b-d)/(b+c-a-d)$, then

$$E(X,S) = pE(X,X) + (1-p)E(X,Y)$$

$$= pa + (1-p)b = \frac{(ab-ad)+(bc-ab)}{b+c-a-d} = \frac{bc-ad}{b+c-a-d}.$$

Similarly, we have

$$E(Y,S) = pE(Y,X) + (1-p)E(Y,Y) = pc + (1-p)d = \frac{bc-ad}{b+c-a-d}.$$

That is, with p chosen as in **(6)**, $E(X,S) = E(Y,S)$.

The theorem and corollary assume that we are given mixed strategies that are ESSs. Now we finally show that a mixed ESS must exist if no pure ESS exists in a two-strategy game. Let $S = pX + (1-p)Y$ now denote the mixed strategy with p chosen as in **(6)**. We show that S is an ESS.

Let $T = qX + (1-q)Y$ be any strategy other than S, with $0 \le q \le 1$. (If $q = 1$ or $q = 0$, then T is the pure strategy X or Y, respectively.) Given our choice of p, we have $E(X,S) = E(Y,S)$, by **Corollary 1** $E(S,S) = E(X,S) = E(Y,S)$. So just as in **(4)** in the proof of **Theorem 1**, $E(T,S) = E(S,S)$. So to show that S is an ESS, we must show that $E(S,T) > E(T,T)$. But using the payoffs in **Table 4**, we get

$$
\begin{aligned}
E(S,T) &= pqa + p(1-q)b + (1-p)qc + (1-p)(1-q)d \\
&= d + (c-d)q + p[b-d+(a+d-b-c)q]. \quad \text{(7)}
\end{aligned}
$$

Similarly, we may determine $E(T,T)$ by just substituting q for p in **(7)**,

$$E(T,T) = d + (c-d)q + q[b-d+(a+d-b-c)q]. \quad \text{(8)}$$

To show that $E(S,T) > E(T,T)$, it suffices to show that $E(S,T) - E(T,T) > 0$. But by **(7)** and **(8)**, we have

$$E(S,T) - E(T,T) = (p-q)[b-d+(a+d-b-c)q]. \quad \text{(9)}$$

Adding $(a+d-b-c)q = -q(b+c-a-d)$ to **(5)**, we obtain

$$b-d+(a+d-b-c)q = (p-q)(b+c-a-d).$$

Substituting this into **(9)**, we find

$$E(S,T) - E(T,T) = (p-q)^2(b+c-a-d) > 0,$$

where the inequality follows because we have assumed there is no pure ESS, so $c > a$ and $b > d$. Thus, we have shown the following.

Theorem 2 *In a two-person game with payoff matrix*

Player 1	Player 2	
	X	Y
X	a	b
Y	c	d

there is always an ESS. If there is no pure ESS, then $S = pX + (1 - p)Y$ *is a mixed ESS with*

$$p = \frac{b - d}{b + c - a - d}.$$

Let's apply **Theorem 2** to the game of Chicken in **Table 2**. Since there is no pure ESS, there must be a mixed ESS where the proportion of Chicken strategists is

$$p = \frac{b - d}{b + c - a - d} = \frac{-110 - (-1010)}{-110 + 90 - (-10) - (-1010)} = 0.90.$$

That is, there is a mixed ESS in which the Chicken strategy is played 90% of the time and Not Chicken is played 10%. This can happen in a variety of ways: 90% of the population might be "pure" Chickens and 10% "pure" Not Chickens, or each individual might be a mixed strategist playing Chicken 90% and Not Chicken 10%, or individuals might play each of the strategies with varying percentages but the overall play in the population is 90% Chicken and 10% Not Chicken.

Even very simple games like this are found in nature. Eagles often engage in spectacular aerial dogfights in which opponents fly at each other, lock their talons together, and fall in a spiral toward the earth. Just before they reach the treetops, the two eagles let go of each other and pull out of the dive. In these contests, the loser is the eagle that lets go first.

Exercises

3. Revise the game of Chicken so that

- each player now puts $200 into the pot,

- it still costs $10 get the car ready, and

- it still costs $1000 to repair in case of an accident.

a) Fill in the following table of payoffs to Player 1 for each possible contest.

Player 1	Player 2	
	Chicken	Not Chicken
Chicken		
Not Chicken		

b) Determine the ESS for the game.

4. Revise the game of Chicken so that

- each player puts $100 into the pot,

- it still costs $1000 to repair in case of an accident, and

- it does not cost anything to get the car ready.

a) Fill in the following table of payoffs to Player 1 for each possible contest.

Player 1	Player 2	
	Chicken	Not Chicken
Chicken		
Not Chicken		

b) Determine the ESS for the game.

5. Consider the following two payoff matrices. Does either game have a pure ESS? If not, find the mixed ESS. Explain.

Player 1	Player 2	
	A	B
A	0	2
B	−3	1

Player 1	Player 2	
	A	B
A	4	3
B	4	−1

6. In a game of Chicken assume that it costs $10 to get the car ready and $1,000 to repair it. How much money would each player have to put in the pot for there to be a mixed ESS in which the proportion of Chicken strategists in the population is $p = 2/3$?

7. a) Return to the game in **Problem 2**. Assume that $x < y < z$. Find the ESS for this game.

b) What happens to the ESS as z, the cost of repairs, increases?

c) What happens to the ESS as y, the amount bet, increases?

d) What happens to the ESS as x, the cost of preparing the car, increases?

3.   Hawk **and** Dove: **A Discrete Symmetric Contest**

The classic Hawk and Dove game involves two discrete strategies. Those individuals playing a Hawk strategy always fight to injure or kill their opponent. Individuals employing the Dove strategy always display and never escalate the contest to serious fighting. (It is important to remember that these two strategies are being played by contestants of the same species.) If two individuals meet and both adopt the Hawk strategy, at least one will be seriously injured in the contest. Likewise, if two players both adopt the Dove strategy, there is some cost to continued displaying. When one player adopts a Hawk strategy and the other plays Dove, the Hawk wins the contested resource (i.e., food, territory, mates).

We now carry out an analysis of Hawk vs. Dove game. First we list, in general terms, the costs and benefits associated with the various strategies.

Table 5.
The costs and benefits associated with the Hawk and the Dove strategies.

Action	Benefit or Cost (arbitrary units)
Gain resource	v
Lose resource	0
Injury to self	i
Cost to display self	t

Here, v is positive, i and t are nonnegative numbers, and the payoffs are in arbitrary units of fitness. We can calculate the payoffs in the various Hawk and Dove contests as was done in the game of Chicken. Hawk always beats Dove; and when the same two strategies compete, we assume that either player has a 50% chance of winning. Therefore,

$$E(H, H) = \tfrac{1}{2}v - \tfrac{1}{2}i,$$
$$E(H, D) = 1 \cdot v - 0 = v,$$
$$E(D, H) = 0 \cdot v + 1 \cdot 0 = 0,$$
$$E(D, D) = \tfrac{1}{2}(v - t) + \tfrac{1}{2}(-t) = \tfrac{1}{2}v - t.$$

The payoff matrix for the Hawk vs. Dove game is given in **Table 6**.

Table 6.
The general Hawk vs. Dove payoff matrix.

Player 1	Player 2	
	Hawk	Dove
Hawk	$\tfrac{1}{2}v - \tfrac{1}{2}i$	v
Dove	0	$\tfrac{1}{2}v - t$

3.1 Pure ESSs for Hawk and Dove

Recall that Hawk is a pure ESS if either $E(H, H) > E(D, H)$ or both $E(H, H) = E(D, H)$ and $E(H, D) > E(D, D)$. Using the values in **Table 6**, the first condition becomes

$$E(H, H) > E(D, H) \iff \tfrac{1}{2}v - \tfrac{1}{2}i > 0 \iff v > i.$$

The second condition is equivalent to

$$E(H, H) = E(D, H) \iff \tfrac{1}{2}v - \tfrac{1}{2}i = 0 \iff v = i$$

and

$$E(H, D) > E(D, D) \iff v > \tfrac{1}{2}v - t.$$

But this latter condition is always true, since v is positive t is nonnegative. Thus, we conclude that Hawk is a pure ESS whenever $v \geq i$, that is, whenever the value of the resource is at least as great as the cost incurred by injury.

Can Dove ever be an ESS? We just saw that $E(H, D) > E(D, D)$, so Dove can never be a pure ESS. This makes sense, since any population of Doves can easily be invaded by Hawks.

3.2   Mixed ESSs for Hawk and Dove

Common assumptions for this game are that the value of the resource is less than the cost of injury (this is what makes life risky) and that twice the cost of display is less than the value of the resource. That is, we assume that $2t < v < i$. In short, injuries are costly but displaying is inexpensive.

However, if $v < i$, then neither Hawk nor Dove is a pure ESS. But by **Theorem 2**, there must be a mixed ESS. If h is the proportion of the Hawk strategy in such a mixed ESS, then using **Table 6** and **Theorem 2** we get

$$h = \frac{v - \left(\tfrac{1}{2}v - t\right)}{v + 0 - \left(\tfrac{1}{2}v - \tfrac{1}{2}i\right) - \left(\tfrac{1}{2}v - t\right)} = \frac{t + \tfrac{1}{2}v}{t + \tfrac{1}{2}i},$$

or more simply,

$$h = \frac{2t + v}{2t + i}. \tag{10}$$

Thus, when injury costs exceed the value of the resource (i.e., when $i > v$), (10) gives the proportion of the Hawk strategy in a mixed ESS as a function of the payoffs. This makes computing the equilibrium frequency a straightforward matter. For example, if the payoffs are $v = 50$, $i = 100$, and $t = 10$, then

$$h = \frac{2(10) + 50}{2(10) + 100} = \frac{70}{120} = 0.583. \tag{11}$$

The frequency of the Dove strategy will be $d = 1 - h = 1 - 0.583 = 0.417$.

3.3   Two-Strategy Contests in Nature

The Hawk vs. Dove game is undoubtedly an oversimplification of the types of animal conflicts that exist in the wild. Nevertheless these strategies represent two extremes of the possible strategies that might be played by wild animals. The model is used mainly to gain insight on how animal behavior evolves. To use game theory models to predict animal behavior, we need to know the range of possible strategies that could be played and the benefits for each. In practice it is difficult to measure costs and benefits in terms of Darwinian (reproductive) fitness. We can, however, use game theory models to make predictions about animal behavior that can be tested experimentally in the field or laboratory.

For example, in **Exercise 8** you are asked to show that in a Hawk-Dove contest, as the resource value v increases, the proportion of Hawk strategists increases in the population. That is, game theory predicts that escalated contests and potentially costly fighting are selected for only if the winners leave more offspring than losers. Measuring fitness is very difficult, in part because it requires longitudinal data (i.e., data collected over several generations). Bonduriansky and Brooks [1999] carried out such a study on antler flies (*Protopiophila litigata*). Male antler flies compete for oviposition sites on discarded moose antlers. Female antler flies prefer to oviposit on the upward-facing surface of antlers. Male antler flies fight aggressively for access to these main oviposition sites, where females are a high-density resource. Bonduriansky and Brooks [1999] show that male antler flies holding high-quality territories are larger, tend to live longer, and have greater lifetime reproductive success. Winners of escalated contests enjoy a significantly higher frequency of mating and greater lifetime reproductive fitness.

The foraging behavior of bald eagles (*Haliaeetus leucocephalus*) involves multiple strategies to acquire prey. In their wintering grounds in Chilkat Valley, Alaska, bald eagles employ two basic strategies: capturing live or unclaimed prey (hunter) or stealing prey from another eagle (robber) [Hansen 1986]. By placing dead salmon on a gravel bar at intervals of approximately 4 m, Hansen [1986] created artificial food patches. Whenever two eagles interacted over a salmon carcass, Hansen recorded the number and type of display, type and duration of attack, winner and loser, injury status, and degree of hunger for each eagle.

In this two-strategy game, the payoff to the robber is frequency-dependent; the fitness of a robber is higher than that of the hunter when robbers are rare. In other words, if everybody steals, there will be no one to steal from. Game theory predicts that the frequencies of hunter and robber will eventually reach equilibrium (ESS), where the payoffs for both strategies are equal. In Hansen's [1986] study, eagles pirated 58% of the time and hunted 42% but ultimately both types of strategists consumed similar amounts of flesh per unit time via each strategy. The risks of stealing vs. hunting were also equal, because no eagles were injured in Hansen's study (although both behaviors are probably not really risk-free). The foraging tactics of bald eagles at Chilkat Valley appear to be an ESS.

Although displays are commonly used by robbers to steal salmon, bald eagles rarely escalate fighting to the point of injury. According to Maynard Smith and Parker [1976], competitors can use traits such as size, age, or hunger level to predict the winner without having to resort to escalated fighting. Eagles apparently use relative body size to settle disputes without escalated fighting. By careful observations, Hansen [1986] found that the larger of the pair of eagles wins 85% of the disputes.

The value of the resource may also change rapidly under certain conditions. Salmon carcasses are plentiful for short periods of time each year. When fish are plentiful (and there are few eagles in the neighborhood), the value of "owning" a fish is small, because it can easily be replaced. As fish become scarce or the cost of replacing a fish rises, the value of the resource increases. Hansen predicted that escalated fighting should increase when fish carcasses become scarce. In other words, animals should take greater risks when a contested resource becomes more valuable. As predicted, display rates, rates of retaliation of hunters against robbers, and rates of physical contact all increased when food levels were low.

One interesting outcome of Hansen's study was the observation that a relatively constant ESS point is possible in eagle hunter-robber contests despite changes in food resource levels. **Table 7** shows that the proportion of "robbers" in the population remained relatively constant despite varying amounts of food available. The "best" strategy for each eagle may depend more on its size or hunger level than on food level. Smaller (or younger) birds may be more successful as hunters, while larger birds may benefit more by stealing.

Table 7.

The frequency of bald eagle foraging strategies under different food levels during three periods of the winter (taken from Table 7 of Hansen [1986]).

Period	Food Level	Hunter	Robber	% Robbers
17 Nov–9 Dec	High	21	47	69
10 Dec–16 Dec	Low	10	16	62
17 Dec–23 Dec	High	8	20	71
Total		39	83	68

Exercises

8. In **(11)**, we saw that in the Hawk–Dove game, with resource value $v = 50$, injury cost $i = 100$, and display cost $t = 10$, there is a mixed ESS, with the equilibrium proportion of Hawks being $h = 0.583$.

   a) Increase the resource value to $v = 60$. Is there still a mixed ESS? What happens to h?

   b) Increase the resource value to $v = 80$. What happens to h?

   c) Explain in one sentence why this makes biological sense.

9. **a)** Reset the value of v to 50 and leave $t = 10$. Increase the cost of injury to $i = 120$. What is the effect on h compared to the value of h in **(11)**?

   b) Now increase the cost of injury further to $i = 150$. What is the effect on h?

   c) Explain in one sentence why this makes biological sense.

10. **a)** Again set the value of $v = 50$ and $i = 100$. Increase the display cost to $t = 20$. What is the effect on h compared to the value of h in **(11)**?

    b) Now increase the display cost to $t = 30$. What is the effect on h?

    c) Explain in one sentence why this makes biological sense.

11. Let $i = 100$ and $t = 10$. What value of v produces a mixed ESS with a population with 50% Hawks and 50% Doves?

12. **a)** Assume the resource value is $v = 100$, the injury cost is $i = 120$, and the display cost is $t = 20$. Fill in the values in the payoff matrix below.

Player 1	Player 2	
	Hawk	Dove
Hawk		
Dove		

   b) Why are neither pure strategies ESSs in this game?

   c) Determine the mixed ESS.

   d) With $i = 120$ and $t = 20$, determine the smallest value of v that makes Hawk a pure ESS.

13. **a)** Double the size of all the costs and benefits of the original example in the text. That is, assume a resource value of $v = 100$, an injury cost of $i = 200$, and display cost of $t = 20$. Fill in the values in the payoff matrix below.

Player 1	Player 2	
	Hawk	Dove
Hawk		
Dove		

   b) Show that there is a mixed ESS and determine h, the equilibrium frequency of the Hawk strategy in the mixed ESS.

   c) Compare this to the equilibrium value of h in the original game **(11)**. What is the effect of the doubling?

4.   Three-Strategy Games

4.1   Asymmetries Between Players

As Hansen's [1986] study of eagle foraging behavior demonstrates, the relative size or age of an opponent is important in determining the outcome of the game. Given that one contestant is bigger and stronger, playing Hawk against a smaller weaker opponent surely means the larger animal is less at risk of suffering an injury. A small eagle is therefore unlikely to escalate fighting against a larger or more experienced eagle. For a model to be realistic, we must take these asymmetries into consideration.

There are basically three kinds of asymmetries:

- asymmetries correlated with fighting ability,

- asymmetries correlated with resource value, and

- uncorrelated asymmetries.

When the resource is worth more to one contestant than to the other, or one contestant is more likely to win a fight because it is larger or stronger, then the contestants have asymmetries correlated with resource value or fighting ability, respectively. Uncorrelated asymmetries are unrelated to fighting ability or resource value. For example, being first to arrive at a resource may give the player an advantage over those arriving later, but the advantage is not due to fighting ability or the value of the resource.

4.2   The Hawk, Dove, Bourgeois Game

Suppose that one contestant arrived at a resource such as a nest site, territory, or harem of females and, in the absence of any opponents, took ownership of the resource. In this case, it might pay the owner to fight harder to retain the resource. The owner has information about the asymmetry that the other players do not (i.e., the owner knows that it owns the resource) and can assess which strategy to play.

Building on our previous Hawk and Dove model, we now add a third strategy that assesses and uses ownership information to decide on how to play. Call this assessor strategy Bourgeois. An individual exhibiting the Bourgeois strategy would play Hawk if it were the resource owner and Dove if it were the intruder. (The word "bourgeois" comes from the French term for middle class and is used in opposition to the proletariat or working class. The middle class enjoyed property ownership, which the lower working class did not.)

Again let v denote the value of the resource contested, i the cost of injury, and t the display cost. As earlier, we assume that the cost of injury exceeds the value of the resource and that the value of the resource exceeds twice the cost of displaying, that is, $2t < v < i$. The payoffs in contests involving only

Hawks and Doves remain as in **Table 6**. We assume that Bourgeois has a 50% chance of owning a resource any time that it competes; so in any contest with Bourgeois, there is a 50% chance that it will act like a Hawk (own the resource) and a 50% chance that it will be a Dove (not own the resource). In other words, $B = \frac{1}{2}H + \frac{1}{2}D$. Therefore, the payoffs in contests involving Bourgeois are:

$$E(H, B) = \frac{1}{2}[E(H, H) + E(H, D)] = \frac{1}{2}\left[\left(\frac{v}{2} - \frac{i}{2}\right) + v\right] = \frac{3v}{4} - \frac{i}{4},$$
$$E(D, B) = \frac{1}{2}[E(D, H) + E(D, D)] = \frac{1}{2}\left[0 + \left(\frac{v}{2} - t\right)\right] = \frac{v}{4} - \frac{t}{2},$$
$$E(B, H) = \frac{1}{2}[E(H, H) + E(D, H)] = \frac{1}{2}\left[\left(\frac{v}{2} - \frac{i}{2}\right)\right] = \frac{v}{4} - \frac{i}{4},$$
$$E(B, D) = \frac{1}{2}[E(H, D) + E(D, D)] = \frac{1}{2}\left[v + \left(\frac{v}{2} - t\right)\right] = \frac{3v}{4} - \frac{t}{2},$$
$$E(B, B) = \frac{1}{2}[E(H, D) + E(D, H)] = \frac{1}{2}[v + 0] = \frac{v}{2}.$$

The calculation of $E(B, B)$ requires a bit of explanation. We assume that when one of the B strategists acts like it owns the resource, the other does not. Hence, $E(B, B) = \frac{1}{2}[E(H, D) + E(D, H)]$. Using the payoffs calculated above, the payoff matrix for the Hawk, Dove, Bourgeois game can be found in **Table 8**.

Table 8.

The payoff matrix for the Hawk, Dove, Bourgeois game.

	Hawk	Dove	Bourgeois
Hawk	$\frac{v}{2} - \frac{i}{2}$	v	$\frac{3v}{4} - \frac{i}{4}$
Dove	0	$\frac{v}{2} - t$	$\frac{v}{4} - \frac{t}{2}$
Bourgeois	$\frac{v}{4} - \frac{i}{4}$	$\frac{3v}{4} - \frac{t}{2}$	$\frac{v}{2}$

4.3 Is Bourgeois a Pure ESS?

To determine whether Bourgeois is a pure ESS, we first compare the payoffs of the other pure strategies played against B to how B fares against itself and then compute the payoff of a general mixed strategy played against B. For Hawk vs. Bourgeois,

$$E(B, B) > E(H, B) \iff \frac{v}{2} > \frac{3v}{4} - \frac{i}{4} \iff \frac{i}{4} - \frac{v}{4} > 0 \iff i > v.$$

This last inequality is always true, since we have assumed that $i > v$. For Dove vs. Bourgeois,

$$E(B, B) > E(D, B) \iff \frac{v}{2} > \frac{v}{4} - \frac{t}{2} \iff \frac{v}{4} > -\frac{t}{2},$$

which is always true because v is positive and t is nonnegative.

Now let $T = qH + rD + (1 - q - r)B$ be any mixed strategy so that either q or r or both are not 0. Then using the previous two results, we get

$$E(T, B) = E(qH + rD + (1 - q - r)B, B)$$

$$= qE(H, B) + rE(D, B) + (1 - q - r)E(B, B)$$
$$< qE(B, B) + rE(B, B) + (1 - q - r)E(B, B)$$
$$= E(B, B).$$

Thus, by **Definition 1**, Bourgeois is a pure ESS if $v < i$.

4.4   The Diagonal Rule

The reason why it was so easy to show that Bourgeois is a pure ESS is that in **Table 8** $E(B, B)$ is the largest payoff in the third column. More precisely, a payoff on the diagonal is the greatest element in the column of a payoff matrix. When this is so, the strategy is a pure ESS.

Let's be completely general about this. Suppose that we have an n-strategy game with strategies $X_1, X_2, \ldots, X_n$. Suppose that in the $n \times n$ payoff matrix for this game, the ith diagonal element, $E(X_i, X_i)$ is the largest element in the i-th column. This means that $E(X_i, X_i) > E(X_j, X_i)$ for all $j \neq i$. Let $T \neq X_i$ be any (mixed) strategy. We can write T as a combination of the pure strategies, $T = p_1 X_1 + p_2 X_2 + \cdots + p_n X_n$, where $p_1 + p_2 + \cdots + p_n = 1$ and each $p_i \geq 0$. Then because $E(X_j, X_i) < E(X_i, X_i)$, we have

$$E(T, X_i) = p_1 E(X_1, X_i) + p_2 E(X_2, X_i) + \cdots + p_n E(X_n, X_i)$$
$$< p_1 E(X_i, X_i) + p_2 E(X_i, X_i) + \cdots + p_n E(X_i, X_i)$$
$$= (p_1 + p_2 + \cdots + p_n) E(X_i, X_i)$$
$$= E(X_i, X_i).$$

So $E(T, X_i) < E(X_i, X_i)$. Thus, we have proven the following result.

Theorem 3 *The Diagonal Rule. In an n-strategy game with pure strategies X_1, $X_2, \ldots, X_n$, if $E(X_i, X_i) > E(X_j, X_i)$ for all $j \neq i$, then X_i is a pure ESS.*

Exercises

14. a) Let the resource value be $v = 50$, the injury cost $i = 100$, and the display cost $t = 10$. Determine the payoff matrix for the Bourgeois game.

   b) Verify that the Bourgeois strategy is ESS.

15. Use the general payoff matrix for the Bourgeois game to answer the following questions.

   a) Suppose that the cost of display is $t = 10$, as usual. What value of v makes $E(D, D) = E(B, D)$?

   b) Does such a value of v make biological sense?

   c) Suppose that the cost of display is $t = 10$ and the value of the resource is $v = 50$. What injury cost i makes $E(H, B) = E(D, B)$?

   d) Does such a value of i make biological sense?

16. a) Suppose we set the display cost to $t = 0$ but leave $v = 50$ and $i = 100$ unchanged. What is the payoff matrix?

b) Because the display cost is 0, we might expect Doves to fare somewhat better? Is Dove an ESS?

17. a) Let $v = 100$ and $i = 100$ and let the display cost $t = 10$. What assumptions about the payoffs is no longer valid?

b) Is there a pure ESS? If so, what is it?

18. a) Can Hawk ever be a pure ESS? We might think so if the injury cost is not too large relative to the value of the resource. Use **Table 8** to show that $E(H, H) > E(B, H)$ whenever $v > i$.

b) Set the value of v to 120. Leave the cost of injury at $i = 100$ (so $v > i$) and the display cost at $t = 10$. Is Hawk a pure ESS?

4.5 Bully: A More Complicated Three-Strategy Game

We now consider a second game involving assessment by players. A Bully strategist first plays Hawk, but after a brief evaluation period it then plays opposite to its opponent's strategy. In other, words, if a Bully's opponent initially plays Hawk, then the Bully soon backs down and plays Dove. If a Bully's opponent initially plays Dove, then the Bully seizes the advantage and continues to play Hawk. This means that when two Bullies meet, they will both eventually adopt the Dove strategy in the contest. The payoffs in contests involving a Bully are determined by the final strategy that the Bully adopts. The payoff matrix is given in **Table 9**.

Table 9.

The payoff matrix for the Hawk–Dove–Bully game.

Player 1	Player 2		
	Hawk	Dove	Bully
Hawk	$\frac{1}{2}v - \frac{1}{2}i$	v	v
Dove	0	$\frac{1}{2}v - t$	0
Bully	0	v	$\frac{1}{2}v - t$

The advantage of the Bully strategy is that, like Hawk, it always beats Dove but does not incur the injury cost of Hawk in contests with Hawk. A Bully fares better than a Dove in all contests except those with Hawks where the two strategies fare equally well.

Are any of the pure strategies an ESS? Assume as before that the cost of injury exceeds the value of the resource and that twice the cost of displaying is less than the value of the resource, $2t < v < i$. Then Hawk is not an ESS, because $E(H, H) = \frac{1}{2}v - \frac{1}{2}i < 0 = E(D, H)$. Dove is not an ESS because $E(D, D) = \frac{1}{2}v - t < v = E(H, D)$. Similarly, Bully is not an ESS because $E(B, B) = \frac{1}{2}v - t < v = E(H, B)$.

Is there a mixed ESS? Suppose that $S = p_1 H + p_2 D + p_3 B$ were a mixed ESS with $p_1 + p_2 + p_3 = 1$. A mixed ESS can be composed of all three strategies (no p_i is 0) or a combination of any two (exactly one p_i is 0). By **Corollary 1**[1], any pure strategies that do appear in a mixed ESS must all have the same payoff when played against S.

We now show that no mixed ESS S can have both Bully and Dove as component strategies. For if S did, then both $p_2 > 0$ and $p_3 > 0$ and, consequently,

$$
\begin{aligned}
E(B, S) &= p_1 E(B, H) + p_2 E(B, D) + p_3 E(B, B) \\
&= p_1 \cdot 0 + p_2 v + p_3 \left(\tfrac{1}{2}v - t\right) \\
&> p_1 \cdot 0 + p_2 \left(\tfrac{1}{2}v - t\right) + p_3 \cdot 0 \\
&= p_1 E(D, H) + p_2 E(D, D) + p_3 E(D, B) \\
&= E(D, S).
\end{aligned}
$$

But this conclusion contradicts the fact that $E(B, S) = E(D, S)$ if both Bully and Dove are part of the mixed ESS. This means that Bully and Dove cannot both be part of any mixed ESS. So any ESS must consist of a mixture of Hawks and Bullies or Hawks and Doves.

Next we show that there is no mixed ESS composed of only Hawks and Doves. Suppose that $S = pH + (1 - p)D$ were an ESS for some p with $0 < p < 1$. Then, by **Theorem 1** $E(S, S) = E(D, S)$. But

$$
E(D, S) = E(D, pH + (1-p)D) = pE(D, H) + (1-p)E(B, D) = 0 + (1-p)(\tfrac{v}{2} - t),
$$

while

$$
E(B, S) = E(B, pH + (1 - p)D) = pE(B, H) + (1 - p)E(B, D) = 0 + (1 - p)v.
$$

Consequently, $E(B, S) > E(D, S) = E(S, S)$, so S cannot be an ESS. This makes biological sense, because Bullies outplay Doves, so Doves should not be part of an ESS.

Is a mixed ESS of Hawks and Bullies possible? Suppose that $S = pH + (1 - p)B$ were an ESS for some p with $0 < p < 1$. By **Theorem 1**, we must have $E(H, S) = E(B, S)$. But

$$
E(H, S) = E(H, pH + (1 - p)B) = p\left(\tfrac{v}{2} - \tfrac{i}{2}\right) + (1 - p)v = v - \tfrac{1}{2}pv - \tfrac{1}{2}pi
$$

and

$$
E(B, S) = E(B, pH + (1 - p)B) = (1 - p)\left(\tfrac{v}{2} - t\right) = \tfrac{1}{2}v - t - \tfrac{1}{2}pv + pt.
$$

---

[1]**Theorem 1** and **Corollary 1** are easily generalized to n-person games and mixed strategies composed of more than two pure strategies.

If $E(H, S) = E(B, S)$, then

$$v - \tfrac{1}{2}pv - \tfrac{1}{2}pi = \tfrac{1}{2}v - t - \tfrac{1}{2}pv + pt.$$

Collecting all the p-terms together yields

$$p\left(t + \tfrac{1}{2}i\right) = t + \tfrac{1}{2}v.$$

Solving for p, we find

$$p = \frac{t + \tfrac{1}{2}v}{t + \tfrac{1}{2}i},$$

or more simply,

$$p = \frac{2t + v}{2t + i}. \tag{12}$$

The proportion p of Hawks in this mixed strategy is the same as in the Hawk-Dove game (see **(10)**). This makes biological sense. Without any Doves present, the Bullies act effectively like Doves. In every encounter (whether with a Hawk or another Bully), they will bluff first and then, having encountered Hawk-like behavior, subsequently adopt the Dove strategy.

We must still show that $S = pH + (1 - p)B$ with p as in **(12)** is an ESS. By **Corollary 1** $E(S, S) = E(H, S) = E(B, S)$. Also notice that since Doves lose to Hawks and Bullies,

$$E(D, S) = E(D, pH + (1 - p)B) = pE(D, H) + (1 - p)E(D, B) = 0. \tag{13}$$

Now let $T = qH + rD + (1 - q - r)B$ be any mixed strategy. We must show that $E(S, S) > E(T, S)$ or $E(S, S) = E(T, S)$ and $E(S, T) > E(T, T)$. Using **Theorem 1** and **(13)**, we have

$$\begin{aligned}
E(T, S) &= E(qH + rD + (1 - q - r)B, S) \\
&= qE(H, S) + rE(D, S) + (1 - q - r)E(B, S) \\
&= qE(S, S) + 0 + (1 - q - r)E(S, S) \\
&= (1 - r)E(S, S) \\
&\leq E(S, S). \tag{14}
\end{aligned}$$

Thus, if $r > 0$ (i.e., Dove is part of the mixed strategy T), then $E(S, S) > (1 - r)E(S, S) = E(T, S)$. However, if T is a mixture of Hawk and Bully only, that is, if $r = 0$ so that $T = qH + (1 - q)B$, then $E(S, S) = E(T, S)$.

So the problem has been reduced to a two-strategy game, Hawk and Bully. But as we observed earlier, without any Doves present, Bullies are reduced to Dove-like behavior. The Hawk and Bully payoffs are given in **Table 10**.

Since the entries in **Table 10** are identical to those in the Hawk-Dove game in **Table 6**, so is the mixed ESS but with Bullies replacing the Doves: $S = pH + (1 - p)B$ with $p = 2t + v/(2t + i)$.

Table 10.

The reduced payoff matrix for `Hawk` vs. `Bully`.

Player 1	Player 2	
	`Hawk`	`Bully`
`Hawk`	$\frac{1}{2}v - \frac{1}{2}i$	v
`Bully`	0	$\frac{1}{2}v - t$

Exercises

19. Consider a new strategy `Dove`. A `Retaliator` plays in a way exactly opposite to a `Bully`. It starts out playing `Dove` and then after a brief evaluation period it adopts the strategy of its opponent. Fill in the payoff table (v, i, and t have their usual meanings) for a `Bully` vs. `Retaliator` game and find the ESS(s) assuming $2t < v < i$. Remember: `Bully` starts out playing `Hawk`.

20. Write the payoff matrix and find the ESS for a `Bully` vs. `Bourgeois` game.

21. Write the payoff matrix and find the ESS for a `Retaliator` vs. `Bourgeois` game.

22. Write out the payoff matrix for a `Dove`, `Bully`, `Retaliator` game. Show that there is only one pure ESS.

23. a) Write out the payoff matrix for the four strategy game involving `Hawk`, `Dove`, `Bully`, and `Retaliator`. Is there a pure ESS?

b) Show that the mixed strategy $S = pH + (1 - p)B$ with p chosen as in **(6)** is an ESS for this new game. Hint: Mimic the ideas following **(12)**.

5.   Asymmetries

All the models described so far assume that all `Hawks` are equally matched (symmetric), as are all `Doves` and `Bourgeois`. In the real world, this is rarely the case. Instead, contestants usually vary in one or more qualities that may have an effect on the outcome of the interaction; they have asymmetries in behavior. Given that asymmetries exist, it would make sense if the contestants could assess these asymmetries in some way and adjust their behavior to the particular situation. Play `Dove` when, after evaluating the asymmetries, you decide the opponent is more likely to win, and play `Hawk` when you decide you have the advantage. The particular asymmetry could be differences in body size, age, ownership of the disputed resource, etc. Cases where opponents can adjust the strategy they play depending on the circumstances are referred to as *conditional strategies*. Conditional strategies can be more successful than the mixed ESS of a randomly played `Hawk-Dove` game.

5.1   Asymmetries in Resource Values

A given resource may be perceived differently by two contestants. A female may be perceived as more valuable if she has not already been inseminated [Austad 1983]. Territories may be more valuable to a resident who has already learned the location of important food patches or nest sites than to an intruder [Beletsky and Orians 1987]. Likewise, an animal that is starving places a higher value on a given food item than an animal that has recently fed. As a result, the starving animal might be more willing to risk injury by adopting a Hawk strategy even in the face of a larger opponent. This prediction has been verified experimentally for common shrews (*Sorex araneus*) [Barnard and Brown 1984]. Shrews have very high metabolic rates and consequently must consume relatively large quantities of food per day, making food a valuable resource. Prior to each experiment, shrews were divided into two groups, those that received a high-density food supply and those that received a low density food supply. After two days, one "high-density shrew" and one "low-density shrew" were placed together in an observation arena where aggressive interactions were scored. In the second stage of the trial the diets of the shrews were reversed: "Low-density shrews" were given access to high-density food resources, "high-density shrews" foraged where prey were at low density, and the experiments were repeated (see **Table 11**.) In both stages, shrews experiencing lower food density won the majority of the interactions. Presumably, hungrier shrews are more willing to risk injury to secure the rights to a disputed food resource.

Table 11.
Percentage of interactions won (number of contests).
Adapted from Barnard and Brow [1984], Table 1.

Prior resource	Density	
experience	Low	High
Stage 1	85.8 (231.3)	14.2 (37.6)
Stage 2	20.4 (45.5)	79.6 (175.5)

Male damselflies (*Calopteryx maculata*) defend territories consisting of small patches of emergent vegetation along the water's edge. Females come to these patches to lay eggs on the vegetation thereby providing the territory holder with an opportunity to mate. Males vigorously defend their small territories from other males by chases lasting from a few seconds to over an hour. Typically, the territory holder wins the contest. When territory holders are away feeding or chasing an intruder, the territory may sometimes be taken over by another male. When the original owner returns, territory ownership is confused and an escalated contest results [Waage 1988; Marden and Waage 1990]. The fact that territory owners tend to defeat intruders in disputes might mean that a Bourgeois strategy is being played. Alternatively, it might mean that territory owners have more to gain and are therefore prepared to fight harder or longer. If the territory is rich in high quality food patches or provides better access to mating opportunities, then the territory has a higher resource value to the owner, because the owner has more knowledge of the territory's characteristics.

In cases where the value of the resource exceeds the cost of injury, we would predict Hawk-like strategies to evolve. Fierce fighting results because losers might fail to pass on any genes to future generations. When the resource value is lower or the injury cost is very high, we would predict that a Bourgeois strategy would evolve to settle disputes.

Male elephant seals (*Mirounga angustirostrus* and *M. leonina*) congregate on certain beaches each year to breed. Large, dominant males arrive first and set up territories on the beach. As females arrive on the breeding beaches, the males sequester the females into harems that are vigorously defended. All matings are performed by these harem masters who defend their females from the many late arriving bachelor males that live at the edge of the sea. Fights between elephant seal bulls are brutal and bloody. The females in the harem benefit by mating only with the largest and strongest bulls. The harem masters risk serious injury and expend so much energy defending their females that they usually retain their harems for only a year or two before dying. As a result, a harem master's entire reproductive success may depend on retaining the harem for a single year [Le Boeuf 1974].

Game theory predicts that as the value of the resource increases, contestants will be more likely to escalate the battle. In the case of male elephant seals, the value of retaining harem ownership is very high: their one chance to breed. For the harem master, a major portion of his lifetime reproductive success is at stake and he has nothing to gain by retreating. Under conditions of extremely high resource values, interaction strategies leading to serious injury or even death can evolve.

5.2   Asymmetries in Fighting Ability

Even if the resource does have equal value to both contestants, not all players are created equal. Often one contestant is larger, heavier, has larger weapons, or has more experience fighting. In these cases, the contestant is said to have greater "resource holding power" (RHP). Rivals would increase their own long-term fitness by accurately assessing their opponent's RHP and adjusting their behavior accordingly. Male deer, for example, assess the size of their opponent's antlers, and male elk assess their opponent's vigor by the duration and intensity of roaring contests [Clutton-Brock and Albon 1979].

In situations where animals can assess the resource holding potential of a rival, the Assessor strategy may prevail [Maynard Smith 1982]. If a contestant's RHP is greater than its opponent's, adopt the Hawk strategy; but if its RHP is less than its opponent's, play Dove. This is an example of a conditional strategy.

Strategies such as Bourgeois and Bully are essentially Assessor strategies, because the decision to escalate or retreat is based on an assessment of the opponent or its strategy. How can opponents assess one another? Perhaps the most common method of assessment is based on differences in body size or strength between opponents. If an `Assessor` can accurately evaluate its relative size or strength, it is likely to fare well against a `Hawk` or `Dove` strategy, because it avoids paying fighting and injury costs when it is likely to lose the contest.

Body size can be evaluated in a number of ways. Several species of frogs and toads evaluate the relative size of an opponent by the frequency range of the opponent's calls [Davies and Halliday 1978]. The physics of sound production make it energetically less expensive for larger toads to produce low-frequency calls. Therefore, call frequency is inversely proportional to caller body size and can be readily evaluated by an opponent without direct observation. Male toads (*Bufo bufo*) fight over gravid females. Davies and Halliday [1978] allowed both large males and small males to amplex (mount and hold onto a female) with gravid females. Each male was temporarily muted and a series of either high or low-frequency calls was played back via a small speaker near the amplexed pair. As predicted, when a male of intermediate size was introduced, it more readily attacked the amplexed male when high-frequency croaks were played, especially if the defender was also smaller in body size. Only when the defender and the attacker were of similar body size did fights escalate. These experiments suggest that male *Bufo* can use a combination of visual and auditory cues to assess the fighting ability of an opponent.

In some cases, relative fighting ability is not related to body size. Briffa and Elwood [2000] studied fighting ability in hermit crabs. Hermit crabs fight by rapping their shells against those of an opponent in an attempt to evict the opponent from its shell. Once evicted, the aggressor exchanges its own shell for the newly vacant shell. The information on fighting ability is conveyed by the force and rate of shell rapping. Briffa and Elwood [2000] experimentally reduced the force with which an aggressor could rap on an opponent's shell by painting the aggressor's shell with a rubberized material. Hermit crabs in rubberized shells were less likely to evict (i.e., win) opponents from their shells. Presumably the force and rate of shell rapping is a more accurate signal of fighting ability than shell size, because small hermit crabs often occupy large shells and large crabs may occupy small shells.

Visual cues are often used to assess relative fighting ability. As previously described for eagles and toads, overall body size is a common method of assessing fighting ability. In other cases, dominant or more aggressive individuals often display badges of status. Many species of lizards display brightly colored throat patches (called *dewlaps*) when confronted by an opponent. Males with dewlaps of specific colors or patterns are significantly more likely to win disputes. Male house sparrows (*Passer domesticus*) exhibit a great deal of variation in the size of the dark patch of feathers on the throat. Males with the largest throat patches are socially dominant to other males, and these "badges" of status serve as effective deterrents.

6.   Application: Musth in African Elephants

Female African elephants (*Loxodonta africana*) are social animals and often live in kinship groups dominated by a matriarchal female. Adolescent and adult males are excluded from these female groups, and become solitary or form loose bachelor herds. Sexually active males (usually older than 25 years of age) seek out females that are in estrus (sexually receptive). Some proportion of adult males experience episodes of *musth*, a period of heightened aggression and sexual activity brought on by elevated levels of testosterone in the blood [Vaughan, et al. 1999]. The frequency of musth and its duration is related to the age of the male. When a male enters a period of musth, it advertises its increased aggressiveness with secretions from a gland near the eye, vocalizations, and increased urine-marking [Poole 1989]. These cues should be relatively easy for other males to assess. Musth males may be signaling their intention to fight (asymmetry in fighting ability or motivation) and that they place a higher value on receptive females than non-musth males (asymmetry in resource value).

During aggressive encounters between sexually active but non-musth males, body size usually determines the outcome. In 86% of the interactions between non-musth males and those in musth, however, the musth male won the contest regardless of size. When both males are in musth, fights often escalate and can result in serious injury and even death [Poole 1989]. Escalated fights can last for several hours during the heat of the day, resulting in significant thermoregulatory costs for both contestants. In addition, musth males spend less time feeding and their overall condition over the 2–4 month musth period frequently deteriorates. The benefits to the winner are measured in terms of evolutionary fitness as the number of offspring sired per year.

According to Poole [1989], at Amosell National Park in Kenya there are approximately 30 estrus females available during the wet season and 15 during the dry season. These females represent the available resources or benefits (see **Table 12**). High-ranking males (H) guard and mate any receptive females they encounter, but lower-ranking males (L) can mate with only those females that are not being guarded by a higher-ranking male.

The costs of securing and defending females were estimated by Poole [1989] by using several methods to score the physical condition of each male before and after the breeding season. Physiological costs are higher during the dry season, because elephants must travel farther to water. Further, the cost of fighting is higher for lower-ranking males, because they are more likely to meet males of equal or higher rank than are higher-ranking males. Moreover, since reproductive success rises rapidly later in life, small young males have more to lose in reproductive success if they are injured or killed in a fight. Lastly, the physiological costs prevent males from being in musth continuously, which leads to three possible strategies for any higher- or lower-ranking male:

Table 12.
Costs, benefits, and payoffs for high- and lower-ranking males.
Adapted from Poole [1989], Tables III–IV.

	Benefits		
	Wet	Dry	Full Yr
All males	30	15	45
	Condition Costs		
All males	10	15	25
	Fighting Costs		
Higher-ranking males	5	5	10
Lower-ranking males	10	10	20

1. Come into musth each year but only during the wet season (W).

2. Come into musth each year but only during the dry season (D).

3. Stay in musth for a full year, but only come into musth every other year (F).

Poole was therefore able to assign qualitative scores to both fighting and condition costs for each male for each of these strategies (see **Table 12**).

A game theory model of this general situation has six different types of players, whose strategies consist of all possible dominance (H or L) and musth season (W, D, or F) combinations. For example, HW represents the strategy of a higher-ranking male coming into musth only in the wet season while LF represents the strategy of a lower-ranking male coming into musth for a full year but only in alternate years.

6.1 The Payoff Matrix

We now determine the entries in the payoff matrix for this six strategy game. For the sake of simplicity and clarity, we make the following assumptions.

1. Benefits consist of the total number of females available during the particular period.

2. If two elephants in musth compete, the higher-ranking male always wins; if both are the same rank, each has a 50% chance of winning.

3. Payoffs are calculated for a two-year period to accommodate the F-strategy: in musth for a full-year in alternate years.

4. In a contest between one elephant adopting a wet season musth strategy and the other a dry season musth strategy, no fighting occurs, so no fighting costs are incurred.

5. In a contest between one elephant adopting a wet (dry) season musth strategy and the other a full-year musth strategy, fighting and its costs occur only in one of the wet (dry) seasons during the two year period.

6. In any one year, only half of the F-strategists are actually in musth. Consequently, in a contest between two F-strategists, there is a 50% chance that they will be in phase (both in musth and out of musth in the same years), so fighting will occur in one of the two years. There is also a 50% chance that they will be out of phase (one in musth and one out of musth in the same years), so fighting will not occur.

Table 13.

Payoff matrix for different strategies. Key: high-ranking male (H), low-ranking male (L), in musth wet season (W), in musth dry season (D), in musth full year in alternate years (F).

	HW	HD	HF	LW	LD	LF
HW	0	40	20	30	40	35
HD	0	−25	−12.5	0	−10	−5
HF	0	7.5	3.75	15	15	15
LW	−40	40	0	−10	40	15
LD	0	−50	−25	0	−35	−17.5
LF	−20	−5	−12.5	−5	2.5	− 1.25

We won't go through all 36 payoff calculations in **Table 13**, but here are a few. You should try to check the values in the rest of the table.

*   For $E(HW, HW)$, there's a 50% chance of winning a wet season benefit of 30 minus the condition and fighting costs for the wet season (all times 2 years).

$$E(HW, HW) = 2[(0.5)30 - 10 - 5] = 0.$$

*   For $E(HW, HD)$, there's a 100% chance of winning a wet season benefit of 30 minus the condition for the wet season; no fighting cost is incurred (all times 2 years).

$$E(HW, HD) = 2(30 - 10) = 40.$$

*   For $E(HW, HF)$, from the point of view of HW, half of the time, the HF is in musth (so fighting occurs with each having a 50% chance of victory)—this is equivalent to $E(HW, HW)$. The other half of the time no fight occurs since HF is not in musth: there's a 100% chance of HW winning—this is equivalent to $E(HW, HD)$.

$$E(HW, HF) = (0.5)[E(HW, HW) + E(HW, HD)] = (0.5)(0 + 40) = 20.$$

*   For $E(HW, LW)$, there's a 100% chance of winning a wet season benefit of 30 minus the condition and fighting costs for the wet season (all times 2 years).

$$E(HW, LW) = 2(30 - 10 - 5) = 30.$$

Notice that $E(HW, LD) = E(HW, HD) = 40$ since no fighting occurs. $E(HW, LF)$ is the average of $E(HW, LW)$ and $E(HW, LD)$, so $E(HW, LF) = 35$. Payoffs for HD are similarly calculated.

- Full-year alternate-year payoffs are more complicated to calculate. For $E(HF, HW)$, from the point of view of HF in its musth year, fighting occurs in the wet season with a 50% chance of victory; no fighting occurs in the dry season (HF receiving the dry season benefit). HF incurs full-year condition costs, but only wet season fighting costs. In the non-musth year, no fighting occurs; there are no benefits and no costs.

$$E(HF, HW) = (0.5)30 + 15 - 25 - 5 = 0.$$

Similarly,
$$E(HF, HD) = 30 + (0.5)15 - 25 - 5 = 7.5.$$

- For $E(HF, HF)$, in its musth year, half the HFs encountered are also in musth, fighting occurs each with an equal chance of victory; the other HFs encountered are not in musth, no fighting occurs and the benefits accrue to the first HF. There are condition costs for the entire year. In its non-musth year, there are no benefits or costs.

$$E(HF, HF) = (0.5)[(0.5)45 - 10] + (0.5)45 - 25 = 3.75.$$

- For $E(HF, LW)$, in its musth year, fighting occurs in the wet season with a 100% chance of victory, no fighting occurs in the dry season (HF receiving the dry season benefit). HF incurs full-year condition costs, but only wet season fighting costs. In the non-musth year, no fighting occurs; there are no benefits and no costs.

$$E(HF, LW) = 30 + 15 - 25 - 5 = 15.$$

- Similarly, $E(HF, LD) = 30 + 15 - 25 - 5 = 15.$

- $E(HF, LF)$, is similar to $E(HF, HF)$, except that when fighting occurs, HF is always the winner.

$$E(HF, LF) = (0.5)[45 - 10] + (0.5)45 - 25 = 45 - 5 - 25 = 15.$$

- For $E(LW, HW)$, there's no chance of winning a wet season benefit while incurring the condition and fighting costs for the wet season (all times 2 years).
$$E(LW, HW) = 2[(0.0)30 - 10 - 10] = -40.$$

- For $E(LW, LW)$, there's a 50% chance of winning a wet season benefit with condition and fighting costs for the wet season (all times 2 years).

$$E(LW, LW) = 2[(0.5)30 - 10 - 10] = -10.$$

- For $E(LW, LF)$, from the point of view of LW, half of the time LF is in musth (so fighting occurs with each having a 50% chance of victory)—this is equivalent to $E(LW, LW)$, the other half of the time no fighting occurs, since LF is not in musth there's a 100% chance of winning—this is equivalent to $E(LW, LD)$.

$$E(LW, LF) = (0.5)[E(LW, LW) + E(LW, LD)] = (0.5)(-10 + 40) = 15.$$

- For $E(LD, HW)$, there's no fighting cost and the dry season benefit accrues to LD.

$$E(LD, HW) = 2(15 - 15) = 0.$$

- For $E(LD, HF)$, when HF is in musth there's fighting and no victory for LD, when HF is not in musth there's no fighting cost and the dry season benefit accrues to LD.

$$E(LD, HF) = 2(0.5)[(-15 - 10) + (15 - 15)] = -25.$$

- For $E(LF, HF)$, in the year when LF is in musth, half the HFs encountered are in musth, there's fighting and no victory for LF, the other half of the time there's no fighting cost and the full season benefit accrues to LF. When LF is not in musth, there are no benefits or costs.

$$E(LF, HF) = (0.5)(-25 - 20) + (0.5)(45 - 25) = -12.5.$$

- For $E(LF, LW)$, in its musth year, fighting occurs in the wet season with a 50% chance of victory, no fighting occurs in the dry season (LF receiving the dry season benefit). LF incurs full-year condition costs, but only wet season fighting costs. In the non-musth year, no fighting occurs; there are no benefits and no costs.

$$E(LF, LW) = (0.5)30 + 15 - 25 - 10 = -5.$$

- $E(LF, LF)$ is similar to $E(HF, HF)$, except the fighting costs are higher.

$$E(LF, LF) = (0.5)[(0.5)45 - 20] + (0.5)45 - 25 = -1.25.$$

6.2   Strategies for High-ranking and for Low-ranking Males

From **Table 13**, we see that for high-ranking males, HW is a pure ESS, because for any pure strategy X we have $E(HW, HW) \geq E(X, HW)$, and whenever $E(HW, HW) = E(X, HW)$, then $E(HW, X) \geq E(X, X)$. Thus, high-ranking males should come into musth only during the wet season.

But if all high-ranking males come into musth in the wet season, then this means that there will be no females available to low-ranking males during the

Table 14.

Costs, benefits, and payoffs for lower-ranking males under the assumption that all higher-ranking males come into musth in the wet season. Adapted from Poole [1989], Tables III–IV.

	Wet	Dry	Full Yr
Benefits	0	15	15
Condition Costs	10	15	25
Fighting Costs	10	10	20

wet season but there will be 15 available during the dry season and, consequently, only 15 available over the full year. That is, **Table 13** is now modified as follows.

The costs and benefits in **Table 14** may be used to construct a revised payoff matrix for lower-ranked elephants. This has been done in in **Table 15**. For example, for $E(LF, LW)$, in its musth year, fighting occurs in the wet season with a 50% chance of victory, no fighting occurs in the dry season (LF receiving the dry season benefit). LF incurs full-year condition costs, but only wet season fighting costs. In the non-musth year, no fighting occurs; there are no benefits and no costs.

$$E(LF, LW) = (0.5)0 + 15 - 25 - 10 = -20.$$

The other entries are calculated similarly.

Table 15.

Two-year payoff matrix for different strategies under the assumption that all high-ranking males come into musth in the wet season.

	HW	LW	LD	LF
LW	−40	−40	−20	−30
LD	0	0	−35	−17.5
LF	−20	−20	−27.5	−23.75

In Garrison Keillor's fictional town of Lake Wobegon, "All the children are above average" [Keillor 1974]. The same is not true for male Amboseli elephants: Some are higher ranked and some are lower ranked. Given that the optimal strategy for the higher-ranking males is to come into musth annually only in the wet season, what is the best strategy for the lower-ranking males? We show that LD is their optimal response under some mild assumptions.

Depending on how you think about it, this might seem to be the "obvious" strategy or it might seem contradictory. It might be obvious, because lower-ranked males cannot win contests with higher-ranked males. Since the higher-ranked males are in musth only in the wet season, the lower-ranked males should avoid being in musth during this period. Therefore, they should not adopt LW, or even LF, since the latter strategy has a wet-season component in alternate years. This leaves LD as the only remaining strategy. But if we look back at **Table 15**, we see that LD is actually the *worst* or most costly response to itself. This means that an entire population of LD strategists could be invaded by either LF strategists or LW strategists. The problem here is that the population includes some higher-ranked males and both the LF and LW strategies are very costly in contests with HW elephants, while LD is not. Consequently, we should expect the response of the lower-ranked elephants to depend on the number of higher-ranked males in the population

More precisely, let h denote the proportion of the male population consisting of the higher-ranked males. We regard h as some fixed but unknown constant and assume, based on earlier calculations, that these males all adopt strategy HW. Let

$$T = pLW + qLD + rLF \tag{15}$$

denote a general strategy adopted by the lower-ranking males, where p, q, and r may vary as long as each is nonnegative and $h + p + q + r = 1$. In particular, $p + q + r = 1 - h$. Recall from **Definition 1** that if for every strategy $T \neq S$, $E(S, S) > E(T, S)$, then S is an ESS. To determine the optimal strategy for the lower-ranking males, we extend this idea as follows. Given that the high-ranking males use strategy HW, we seek a strategy S for the low-ranking males so that for any strategy T as in **(15)** with $T \neq S$, we have

$$E(S, hHW + S) > E(T, hHW + S), \tag{16}$$

where $S = p'LW + q'LD + r'LF$ with p', q', and r' nonnegative and $h + p' + q' + r' = 1$. In other words, S is the best strategy for lower-ranked elephants to adopt in a world where all the higher-ranked elephants play HW.

We show that the strategy $S = (1 - h)LD$ satisfies **(16)**. That is, lower-ranked elephants should adopt a strategy of coming into musth only in the dry season. Some preliminary calculations simplify the process. Using **Table 15**, we have

$$E(LW, hHW + (1 - h)LD) = hE(LW, HW) + (1 - h)E(LW, LD)$$
$$= -40h - 20(1 - h) = -20 - 20h.$$

Similarly,

$$E(LD, hHW + (1 - h)LD) = 0 - 35(1 - h) = -35 + 35h$$

and

$$E(LF, hHW + (1 - h)LD) = -20h - 27.5(1 - h) = -27.5 + 7.5h.$$

Next, observe that

$$E(LD, hHW + (1 - h)LD) > E(LW, hHW + (1 - h)LD)$$

$$\Longleftrightarrow -35 + 35h > -20 - 20h$$
$$\Longleftrightarrow 55h > 15$$
$$\Longleftrightarrow h > \frac{3}{11}. \tag{17}$$

Similarly,

$$E(LD, hHW + (1 - h)LD) > E(LF, hHW + (1 - h)LD)$$
$$\Longleftrightarrow -35 + 35h > -27.5 + 7.5h$$
$$\Longleftrightarrow 27.5h > 7.5$$
$$\Longleftrightarrow h > \frac{3}{11}. \tag{18}$$

Now assume that $h > 3/11 \approx 0.273$. Biologically, this assumption means that high-ranking elephants compose somewhat more than a quarter of the adult male Amboseli elephant population. If $T \neq hHW + (1 - h)LD$ is any strategy for lower-ranked males, $T = pLW + qLD + rLF$, then, using **(17)**, **(18)**, and the fact that $p + q + r = 1 - h$, we have

$$\begin{aligned}
E(T, hHW + (1 - h)LD) &= E(pLW + qLD + rLF, hHW + (1 - h)LD) \\
&= pE(LW, hHW + (1 - h)LD) \\
&\quad + qE(LD, hHW + (1 - h)LD) \\
&\quad + rE(LF, hHW + (1 - h)LD) \\
&< pE(LD, hHW + (1 - h)LD) \\
&\quad + qE(LD, hHW + (1 - h)LD) \\
&\quad + rE(LD, hHW + (1 - h)LD) \\
&= (p + q + r)E(LD, hHW + (1 - h)LD) \\
&= (1 - h)E(LD, hHW + (1 - h)LD) \\
&= E((1 - h)LD, hHW + (1 - h)LD). \tag{19}
\end{aligned}$$

Thus, we have shown that $S = (1 - h)LD$ satisfies **(16)** and is an uninvadeable strategy if $h > \frac{3}{11}$.

Using the payoffs in **Table 13** and **Table 15**, our model predicts that high-ranking males should come into musth only during the wet season and that lower-ranking males should come into musth only during the dry season. Are these predictions supported for wild Amboseli elephant populations? This is more or less what Poole observed. The high-ranking males came into musth primarily in the wet season and the lower-ranking males (Poole's medium category), came into musth primarily in the dry season. Good! The model seems to describe observed behavior. As Poole [1989] states,

Since fights frequently lead to injury or death, thereby reducing future reproductive potential, elephants should clearly signal, by not being in musth, that they will not fight when the benefits derived from winning are relatively less than they could achieve either at a different time of year or at a later stage of life. Since the number and fighting ability of males in musth changes frequently, males must continuously reassess their role in each asymmetry.

Exercises

24. Suppose that through hunting, all higher-ranked (larger, older) males were removed from the population. Or alternatively, suppose a population of lower-ranked males is relocated to some area (along with females) to reintroduce the species. In either case, there would be only lower-ranked males.

    a) Why are the payoffs for the various strategies in the lower right corner of **Table 13** and not **Table 15**?

    b) Is any pure strategy an ESS?

    c) Is there a mixed strategy ESS?

25. Poole suggests that a more realistic assumption is that when males are in musth, a confrontation between a higher- and lower-ranked male is won 93% of the time by the higher-ranked male rather than 100%. Redo the analysis in this section with that assumption.

7. Answers to the Exercises

1. a) $E(B, A) = -3$.

    b) Both strategies are pure ESSs! $E(A, A) = 0 > -3 = E(B, A)$ and $E(B, B) = 4 > 2 = E(A, B)$.

2.

Player 1	Player 2	
	Chicken	Not Chicken
Chicken	$-x$	$-x - y$
Not Chicken	$y - x$	$-x - z$

3. a)

Player 1	Player 2	
	Chicken	Not Chicken
Chicken	-10	-210
Not Chicken	190	-1010

    b) The mixed ESS has 80% Chickens.

4. a)

Player 1	Player 2	
	Chicken	Not Chicken
Chicken	0	-100
Not Chicken	100	-1000

b) The mixed ESS has 90% Chickens.

5. A is a pure ESS in the first game because $E(A, A) = 0 > -3 = E(B, A)$. A is also pure ESS in the second game because $E(A, A) = E(B, A)$ and $E(A, B) = 3 > -1 = E(B, B)$.

6. Each player would have to put $333.33 into the pot.

7. a) The proportion of Chickens is $(z - y)/z$.

   b) The proportion of Chickens increases as z does.

   c) The proportion of Chickens decreases as y increases.

   d) The proportion of Chickens is independent of the cost to prepare the car.

8. a) Yes, and $h = \frac{2}{3}$.

   b) $h = \frac{5}{6}$.

   c) h increases as v does because it is worth fighting for a more valuable resource given that the cost of injury remains the same.

9. a) $h = \frac{1}{2}$.

   b) $h = \frac{7}{17}$.

   c) h decreases as i increases because it becomes more costly to fight given that the value of the resource remains the same.

10. a) $h = \frac{9}{14}$.

   b) $h = \frac{11}{16}$.

   c) h increases as t does because it becomes more costly to display (to be a Dove) given that the value of the resource and the cost of injury remain the same.

11. $v = 40$.

12. a)

Player 1	Player 2	
	Hawk	Dove
Hawk	-10	100
Dove	0	30

b) Because

$$E(D, H) = 0 > -10 = E(H, H) \quad \text{and} \quad E(H, D) = 100 > E(D, D) = 30.$$

c) $h = 0.875$.

d) $v = 120$.

13. a)

Player 1	Player 2	
	Hawk	Dove
Hawk	−50	100
Dove	0	30

b) $h = 0.583$.

c) h is the same as in the original game.

14. a)

	Hawk	Dove	Bourgeois
Hawk	−25	50	12.5
Dove	0	15	7.5
Bourgeois	−12.5	32.5	25

b) Bourgeois is an ESS by the diagonal rule.

15. a) $v = -20$.

b) This does not make biological sense, since the resource should have a positive value.

c) $i = 120$.

d) Yes, cost of 120 units of fitness makes sense.

16. a)

	Hawk	Dove	Bourgeois
Hawk	−25	50	12.5
Dove	0	25	12.5
Bourgeois	−12.5	37.5	25

b) Bourgeois is still a pure ESS by the diagonal rule.

17. a) We no longer have $v < i$.

b) There is no longer a pure ESS, because even though the first part of **Definition 1** is satisfied by both Hawk and Bourgeois, the second part is not.

18. a) If $v > i$, then $\frac{v}{2} - \frac{i}{2} = \frac{v-i}{2} > 0$. So $E(H,H) = \frac{v-i}{2} > \frac{v-i}{4} = E(H,B)$. Further, $E(H,H) = \frac{v-i}{2} > 0 = E(H,D)$. So by the diagonal rule, Hawk is a pure ESS.

b) Yes, by **a)**.

19. Since Bully starts out Hawk-like and Retaliator starts out Dove-like, the Bully judges the Retaliator to be a Dove and so continues its Hawk-like behavior. The Retaliator does just that—retaliates—and responds with Hawk-like behavior. So Bully-Retaliator contests end up with Hawk-Hawk payoffs.

Player 1	Player 2	
	Bully	Retaliator
Bully	$\frac{1}{2}v - t$	$\frac{1}{2}v - \frac{1}{2}i$
Retaliator	$\frac{1}{2}v - \frac{1}{2}i$	$\frac{1}{2}v - t$

Both Bully and Retaliator are ESSs.

20.

Player 1	Player 2	
	Bully	Bourgeois
Bully	$\frac{1}{2}v - t$	$\frac{1}{2}v$
Bourgeois	$\frac{1}{2}v$	$\frac{1}{2}v$

Bourgeois is a pure ESS.

21.

Player 1	Player 2	
	Retaliator	Bourgeois
Retaliator	$\frac{1}{2}v - t$	$\frac{1}{2}v - \frac{1}{4}i - \frac{1}{2}t$
Bourgeois	$\frac{1}{2}v - \frac{1}{4}i - \frac{1}{2}t$	$\frac{1}{2}v$

Bourgeois is a pure ESS. Retaliator is also a pure ESS, because $i > 2t$ implies $\frac{1}{4}i + \frac{1}{2}t > t$.

22.

	Dove	Bully	Retaliator
Dove	$\frac{1}{2}v - t$	0	$\frac{1}{2}v - t$
Bully	v	$\frac{1}{2}v - t$	$\frac{1}{2}v - \frac{1}{2}i$
Retaliator	$\frac{1}{2}v - t$	$\frac{1}{2}v - \frac{1}{2}i$	$\frac{1}{2}v - t$

Only Bully is an ESS. Retaliator does not satisfy the second part of the definition of an ESS.

23. a)

	Hawk	Dove	Bully	Retaliator
Hawk	$\frac{1}{2}v - \frac{1}{2}i$	v	v	$\frac{1}{2}v - \frac{1}{2}i$
Dove	0	$\frac{1}{2}v - t$	0	$\frac{1}{2}v - t$
Bully	0	v	$\frac{1}{2}v - t$	$\frac{1}{2}v - \frac{1}{2}i$
Retaliator	$\frac{1}{2}v - \frac{1}{2}i$	$\frac{1}{2}v - t$	$\frac{1}{2}v - \frac{1}{2}i$	$\frac{1}{2}v - t$

No strategy is a pure ESS.

b) Hint: Show that $E(R, S) = \frac{1}{2}v - \frac{1}{2}i < E(S, S)$ and that $E(D, S) = 0 < E(S, S)$. Let $T = qH + rD + xB + (1 - q - r - x)R$ and argue as in **(14)**.

24. a) The payoffs in **Table 15** assume that there are no females available to low-ranking males during the wet season because of the presence of high-ranking males in musth at this time. If no high-ranking males are present at all, then the assumptions in **Table 13** are valid viz., females are available to low-ranking males during the wet season.

b) No.

c) Assume that $S = p_1 LW + p_2 LD + p_3 LF$ is a mixed ESS. Using **Theorem 2**, then

$$E(LW, S) = E(LD, S) \implies -10p_1 + 40p_2 + 15p_3 = -35p_2 - 17.5p_3$$
$$\implies -10p_1 + 75p_2 + 32.5p_3 = 0.$$

Using this result and the fact that $p_1 + p_2 + p_3 = 1$, we have a system of two linear equations whose solutions are

$$p_2 = p_1 - \tfrac{13}{17}, \qquad p_3 = -2p_1 + \tfrac{30}{17},$$

where $13/17 \le p_1 \le 15/17$ since each $p_i \ge 0$. A particular solution to this system is $T = \frac{15}{17}LW + \frac{2}{17}LD$. Let $S = p_1 LW + p_2 LD + p_3 LF$ be *any other* solution. Check that

$$E(S, S) = E(T, S) = E(S, T) = E(T, T) = -\tfrac{70}{17}.$$

Now by **Definition 1** neither S nor T is an ESS. Consequently, there are no mixed ESSs.

25. The new payoff matrix, which is the analog to **Table 13**, is

HW	0	40	20	25.8	40	32.9
HD	0	-25	-12.5	0	-12.1	-6.05
HF	0	7.5	3.75	12.9	13.95	13.425
LW	-35.8	40	2.1	-10	40	15
LD	0	-47.9	-23.95	0	-35	-17.5
LF	-17.9	-3.95	-10.925	-5	2.5	-1.25

HW is still a pure ESS. (Why?) So high-ranking males should come into musth only during the wet season. (In fact, this is what Poole observed.) To determine the strategy for lower-ranked males, we use the analog to **Table 15**:

	HW	LW	LD	LF
LW	−35.8	−40	−20	−30
LD	0	0	−35	−17.5
LF	−17.9	−20	−27.5	−23.75

We argue as before. Let h denote the fixed but unknown proportion of the male population consisting of the higher-ranked males. Let $T = pLW + qLD + rLF$ denote a general strategy adopted by the lower-ranking males, where p, q, and r may vary as long as each is nonnegative and $h+p+q+r = 1$. We show that the strategy $S = (1 - h)LD$ still satisfies **(16)**. Lower-ranked elephants should come into musth only in the dry season. Using the analog above to **Table 15**, we get

$$E(LW, hHW + (1 - h)LD) = -35.8h - 20(1 - h) = -20 - 15.8h.$$

Similarly,

$$E(LD, hHW + (1 - h)LD) = -35 + 35h$$

and

$$E(LF, hHW + (1 - h)LD) = -27.5 + 9.6h.$$

Next,

$$E(LD, hHW + (1 - h)LD) > E(LW, hHW + (1 - h)LD)$$
$$\Longleftrightarrow -35 + 35h > -20 - 15.8h$$
$$\Longleftrightarrow 50.8h > 15$$
$$\Longleftrightarrow h > \frac{15}{50.8} \approx 0.2953. \qquad \textbf{(20)}$$

Similarly,

$$E(LD, hHW + (1 - h)LD) > E(LF, hHW + (1 - h)LD)$$
$$\Longleftrightarrow -35 + 35h > -27.5 + 9.6h$$
$$\Longleftrightarrow h > \frac{7.5}{25.4} \approx 0.2953. \qquad \textbf{(21)}$$

Now assume that $h \geq 0.30$, which means that high-ranking elephants compose at least 30% of the adult male Amboseli elephant population. If $T \neq hHW + (1 - h)LD$ is any strategy for lower-ranked males, $T = pLW + qLD + rLF$, then exactly the same argument as in **(19)**, but using **(20)** and **(21)**, shows that $S = (1-h)LD$ satisfies **(16)** and is an uninvadeable strategy.

References

Austad, S.N. 1983. A game theoretical interpretation of male combat in the bowl and doily spider *Frontinella pyramitela*. *Animal Behaviour* 31: 59–73.

Barnard, C.J. and C.A.J. Brown. 1984. A payoff asymmetry in resident-resident disputes between shrews. *Animal Behaviour* 32: 302–304.

Beletsky, L.D., and G.H. Orians. 1987. Territoriality among male red-winged blackbirds II. Removal experiments and site dominance. *Behavioral Ecology and Sociobiology* 20: 339–349.

Bradbury, J.W., and S.L. Vehrencamp. 1998. *Principles of Animal Communication*. Sunderland, MA: Sinauer Associates, Inc.

Bonduriansky, R., and R.J. Brooks. 1999. Why do male antler flies (*Protopiophila litigata*) fight? The role of male combat in the structure of mating aggregations on moose antlers. *Ethology, Ecology, and Evolution* 11: 287–301.

Briffa, M., and R. W. Elwood. 2000. The power of shell rapping influences rates of eviction in hermit crabs. *Behavioral Ecology* 11: 288–293.

Clutton-Brock, T.H., and S.D. Albon. 1979. The roaring of red deer and the evolution of honest advertisement. *Behaviour* 69: 145–170.

Davies, N.B., and T.R. Halliday. 1978. Deep croaks and fighting assessment in toads *Bufo bufo*. *Nature* 274: 683–685.

Hammerstein, P. 1998. What is evolutionary game theory? In *Game Theory and Animal Behavior*, edited by L.A. Dugatkin and H.K. Reeve, 3–15. Oxford: Oxford University Press.

Hansen, A.J. 1986. Fighting behavior in bald eagles: A test of game theory. *Ecology* 67: 787–797.

Keillor, Garrison. 1974. *A Prairie Home Companion*. Weekly radio program, 1974–1987 and subsequently. St. Paul, MN: National Public Radio and subsequently Public Radio International.

Le Boeuf, B.J. 1974. Male-male competition and reproductive success in elephant seals. *American Zoologist* 14: 163–176.

Marden, J.H., and J.K. Waage. 1990. Escalated damselfly territorial contests are energetic wars of attrition. *Animal Behaviour* 39: 954–959.

Maynard Smith, J. 1974. The theory of games and the evolution of animal conflicts. *Journal of Theoretical Biology* 47: 209–221.

_____. 1982. *Evolution and the Theory of Games*. Cambridge: Cambridge University Press.

_____, and G.R. Price. 1973. The logic of animal conflict. *Nature* 246: 15–18.

_____, and G. Parker. 1976. The logic of asymmetric contests. *Animal Behaviour* 24: 159–175.

McCullough, D.R. 1969. The tule elk, its history, behavior, and ecology. *University of California Publications in Zoology* 88: 1–209.

Mesterton-Gibbons, M., and E.S. Adams. 1998. Animal contests as evolutionary games. *American Scientist* 86: 334–341.

Parker, G.A., and D.I. Rubenstein. 1981. Role assessment, reserve strategy, and acquisition of information in asymmetric animal conflicts. *Animal Behaviour* 29: 221–240.

Poole, J.H. 1989. Announcing intent: the aggressive state of musth in African elephants. *Animal Behaviour*, 37: 140–152.

Prestwich, Kenneth N. 1999. Game Theory Website v2.0. www.holycross.edu/departments/biology/kprestwi/behavior/ESS/ESS_index_frmset.html.

Straffin, P.D. 1993. *Game Theory and Strategy*. Washington, DC: Mathematical Association of America.

Vaughan, T.A., J.M. Ryan, and N.J. Czaplewski. 1999. *Mammalogy*. Philadelphia, PA: Saunders College Publishing.

von Neumann, J., and O. Morgenstern. 1944. *Theory of Games and Economic Behaviour*. Princeton, NJ: Princeton University Press.

Waage, J.K. 1988. Confusion over residency and the escalation of damselfly territorial disputes. *Animal Behaviour* 36: 586–595.

About the Authors

Kevin Mitchell received his B.A. in mathematics and philosophy from Bowdoin College and his Ph.D. in mathematics from Brown University. His main areas of interest are hyperbolic tilings, algebraic geometry, and applications of mathematics to environmental science. He has directed four term-abroad programs in Queensland, Australia for Hobart and William Smith Colleges. He is co-author of the recently published text *An Introduction to Biostatistics*.

James Ryan received his B.A. in zoology from SUNY College at Oswego, NY, his M.S. in biological sciences from the University of Michigan, and his Ph.D. from the University of Massachusetts, Amherst. His main areas of interest are functional morphology and ecology of vertebrates and the evolution of vertebrate neural control systems. He is a co-author of *Mammalogy*, a college textbook on mammalian biology, and recently was awarded a grant by the National Geographic Society to  fund an expedition to Bwindi Impenetrable National Park in Uganda to study and film the hero shrew, a rare and unusual small mammal.

This Module was developed by the authors at Hobart and William Smith Colleges for their team-taught interdisciplinary course entitled "Mathematical Models and Biological Systems."

UMAP

Modules in Undergraduate Mathematics and Its Applications

Published in cooperation with

The Society for Industrial and Applied Mathematics,

The Mathematical Association of America,

The National Council of Teachers of Mathematics,

The American Mathematical Association of Two-Year Colleges,

The Institute for Operations Research and the Management Sciences, and

The American Statistical Association.

Module 789

Liouville's Theorem in Dynamics

Paul Atsusi Isihara
Benjamin Noonan

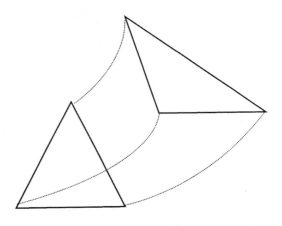

Application Field: Physics and Astronomy

COMAP, Inc., Suite 210, 57 Bedford Street, Lexington, MA 02420 (781) 862–7878

| INTERMODULAR DESCRIPTION SHEET: | UMAP Unit 789 |

TITLE: Liouville's Theorem in Dynamics

AUTHOR: Paul Atsusi Isihara
Dept. of Mathematics
Wheaton College
Wheaton, IL 60187
Paul.A.Isihara@wheaton.edu

Benjamin Noonan
Ben.J.Noonan@wheaton.edu
Nathaniel Stapleton
Nat.J.Stapleton@wheaton.edu

MATHEMATICAL FIELD: Calculus of variations, dynamical systems

APPLICATION FIELD: Physics, astronomy

TARGET AUDIENCE: Undergraduate math majors who have completed a basic calculus and physics sequence.

ABSTRACT: The purpose of this Module is to give an elementary introduction to Liouville's theorem in dynamics. Well-known to physicists, and in particular, fundamental within statistical mechanics, this theorem has many applications, including the focusing of charged particle beams by accelerators and the determination of information about galactic systems. We begin the Module by providing pertinent background material on Euler's equation in the calculus of variations (Section 2) and on Hamiltonian dynamics and phase space (Section 3). We then derive Liouville's theorem in two dimensions (Section 4). An example of three falling balls is given to help visualize an important fact related to Liouville's theorem, namely, that an energy-conserving flow through phase space is incompressible (Section 5). We conclude the Module with further discussion of two of the theorem's applications (Section 6).

PREREQUISITES: A basic knowledge of physics and of multivariable calculus.

UMAP/ILAP Modules 2002–03: Tools for Teaching, 49–67.
©Copyright 2003 by COMAP, Inc. All rights reserved.

COMAP, Inc., Suite 210, 57 Bedford Street, Lexington, MA 02420
(800) 77-COMAP = (800) 772-6627, or (781) 862-7878; http://www.comap.com

Liouville's Theorem in Dynamics

Paul Atsusi Isihara
Dept. of Mathematics
Wheaton College
Wheaton, IL   60187
Paul.A.Isihara@wheaton.edu

Benjamin Noonan
Ben.J.Noonan@wheaton.edu

Nathaniel Stapleton
Nat.J.Stapleton@wheaton.edu

Table of Contents

MODULES AND MONOGRAPHS IN UNDERGRADUATE
MATHEMATICS AND ITS APPLICATIONS (UMAP) PROJECT

The goal of UMAP is to develop, through a community of users and developers, a system of instructional modules in undergraduate mathematics and its applications, to be used to supplement existing courses and from which complete courses may eventually be built.

The Project was guided by a National Advisory Board of mathematicians, scientists, and educators. UMAP was funded by a grant from the National Science Foundation and now is supported by the Consortium for Mathematics and Its Applications (COMAP), Inc., a nonprofit corporation engaged in research and development in mathematics education.

Paul J. Campbell   Editor
Solomon Garfunkel   Executive Director, COMAP

1.   Introduction

We give a concrete introduction to an important theorem in dynamical systems known as *Liouville's theorem*. Well-known to physicists and fundamental in the field of statistical mechanics, this theorem has been widely applied—for instance, to study the focusing of charged particle beams by accelerators and to determine the potential function of the galactic gravitational field from the distribution of stars [Marion and Thornton 1995].

We begin by providing pertinent background on Euler's equation in the calculus of variations (**Section 2**) and on Hamiltonian dynamics and phase space (**Section 3**). We then derive Liouville's theorem in two dimensions (**Section 4**), the more general $2N$-dimensional case being proved in a similar manner. The dynamics of three freely-falling balls helps to visualize an important fact related to Liouville's theorem, namely, the incompressibility of an energy-conserving flow through phase space (**Section 5**). We conclude with further mention of the theorem's applications (**Section 6**).

2.   Euler's Equation from the Calculus of Variations

Hamilton's principle, a fundamental insight in basic classical dynamics, posits that

an object moves in such a way that the time integral of the difference between its kinetic and potential energies is minimized

[Marion and Thornton 1995].

We explain how the calculus of variations is used to minimize such integrals.

First, consider an integral of the form

$$J = \int_{t_1}^{t_2} f\{y(t), \dot{y}(t); t\} \, dt,$$

where y and $\dot{y} = dy/dt$ both depend on t. Our goal is to find a function $y = y_0(t)$ that minimizes J. For example, let us take $f\{y, \dot{y}; t\} = \sqrt{1 + (\dot{y})^2}, t_1 = 0, t_2 = 1$, so that

$$J = \int_0^1 \sqrt{1 + (\dot{y})^2} \, dt.$$

If we specify further that $y(0) = 0$ and $y(1) = 1$, then J gives the length of the graph of a differentiable function $y = f(t)$ that passes through the points $(0, 0)$ and $(1, 1)$. In this case, we know that the function that minimizes J must be $y_0(t) = t$, since the shortest distance between two points is a straight line.

To arrive at this conclusion by means of a calculus-of-variations approach, we begin by setting

$$y = y(\alpha, t) = y_0(t) + \alpha\eta(t).$$

1

Here $\eta(t)$ is an arbitrary differentiable function defined on $t_1 \leq t \leq t_2$ that satisfies

$$\eta(t_1) = \eta(t_2) = 0.$$

For every real number α, the function y agrees with the unknown optimizing function $y_0(t)$ at the endpoints $t = t_1$ and $t = t_2$, but it may vary from $y_0(t)$ in the interval $t_1 < t < t_2$. Note that y and $\dot{y}$ are now functions of both t and α, with $\partial y / \partial \alpha = \eta(t)$ and $\partial \dot{y} / \partial \alpha = \eta'(t)$. The integral J must also depend on α:

$$J = J(\alpha) = \int_{t_1}^{t_2} f\{y(\alpha, t), \dot{y}(\alpha, t); t\} \, dt,$$

and we have

$$
\begin{aligned}
\frac{\partial J}{\partial \alpha} &= \int_{t_1}^{t_2} \frac{\partial}{\partial \alpha} \left[f\{y, \dot{y}; t\} \right] \, dt \\
&= \int_{t_1}^{t_2} \left(\frac{\partial f}{\partial y} \frac{\partial y}{\partial \alpha} + \frac{\partial f}{\partial \dot{y}} \frac{\partial \dot{y}}{\partial \alpha} \right) \, dt \\
&= \int_{t_1}^{t_2} \left(\frac{\partial f}{\partial y} \eta(t) + \frac{\partial f}{\partial \dot{y}} \eta'(t) \right) \, dt.
\end{aligned}
$$

Using integration by parts, we get

$$\frac{\partial J}{\partial \alpha} = \int_{t_1}^{t_2} \frac{\partial f}{\partial y} \eta(t) \, dt + \eta(t) \frac{\partial f}{\partial \dot{y}} \Big|_{t_1}^{t_2} - \int_{t_1}^{t_2} \eta(t) \frac{d}{dt} \left[\frac{\partial f}{\partial \dot{y}} \right] \, dt.$$

Because $\eta(t_1) = \eta(t_2) = 0$, the second term drops out; and after simplification, we have

$$\frac{\partial J}{\partial \alpha} = \int_{t_1}^{t_2} \left(\frac{\partial f}{\partial y} - \frac{d}{dt} \left[\frac{\partial f}{\partial \dot{y}} \right] \right) \eta(t) \, dt.$$

Since J has a minimum at $\alpha = 0$, $\partial J / \partial \alpha$ must equal zero when $\alpha = 0$. Since the function $\eta(t)$ is arbitrary, we must have

$$\frac{\partial f}{\partial y} - \frac{d}{dt} \left[\frac{\partial f}{\partial \dot{y}} \right] = 0. \quad \text{(Euler's equation)}.$$

Euler's equation gives a necessary condition for J to have a minimum.

For the arclength example, $f = \sqrt{1 + (\dot{y})^2}$ and $\partial f / \partial y = 0$, so Euler's equation specifies that

$$\frac{d}{dt} \left[\frac{\partial f}{\partial \dot{y}} \right] = \frac{d}{dt} \left[\frac{\dot{y}}{\sqrt{1 + (\dot{y})^2}} \right] = 0,$$

or

$$\frac{\dot{y}}{\sqrt{1 + (\dot{y})^2}} = c \quad \text{(where } c \text{ is a constant)}.$$

Solving for $\dot{y}$, we find that

$$\dot{y} = \sqrt{\frac{c^2}{1 - c^2}} = \text{constant},$$

and hence y must be linear.

Exercise

1. Use Euler's equation to find the function y that minimizes

$$J = \int_{t_1}^{t_2} \left(\tfrac{1}{2}\dot{y}^2 + y\right) dt,$$

where y satisfies the initial conditions $y(0) = y_0$, $\dot{y}(0) = \dot{y}_0$.

3.   Hamiltonian Dynamics and Phase Space

If an object's motion is constrained to one dimension, the position of the object at time t is given by an ordinary real valued function $y(t)$. In Newtonian dynamics, the motion of the object is analyzed by consideration of the total force acting on the object and Newton's Second Law:

$$\text{Force} = \text{mass} \cdot \text{acceleration} = m\ddot{y}(t).$$

In Hamiltonian dynamics, the object's motion is obtained by consideration of total energy rather than total force. For simplicity, we assume that the object's kinetic energy T is a function of the object's position $y(t)$ and momentum $m\dot{y}$ (the object's mass m is a constant). The object's potential energy, on the other hand, depends only on its position $y(t)$. Thus, we may write

$$T = T(y, \dot{y}), \qquad U = U(y).$$

In particular, we assume that neither T nor U has a direct dependence on time t. Hamilton's principle tells us that the object "moves in such a way that the time integral of the difference between its kinetic and potential energies is minimized" [Marion and Thornton 1995]. In other words, the integral $J = \int_{t_1}^{t_2} L \, dt$, where $L = T - U$, is minimized (L is called the *Lagrangian*). Taking $f\{y, \dot{y}; t\} = L(y, \dot{y}, t) = T(y, \dot{y}) - U(y)$, Euler's equation implies that

$$\left(\frac{\partial T}{\partial y} - \frac{dU}{dy}\right) - \frac{d}{dt}\left[\frac{\partial T}{\partial \dot{y}}\right] = 0,$$

which leads to the equality

$$\frac{\partial T}{\partial y} = \frac{dU}{dy} + \frac{d}{dt}\left[\frac{\partial T}{\partial \dot{y}}\right].$$

3

The total derivative of L indicates how L changes as the object moves through time. We compute this total derivative as follows:

$$
\begin{aligned}
\frac{dL}{dt} &= \frac{\partial T}{\partial y}\dot{y} + \frac{\partial T}{\partial \dot{y}}\ddot{y} - \frac{dU}{dy}\dot{y} \\
&= \left(\frac{dU}{dy} + \frac{d}{dt}\left[\frac{\partial T}{\partial \dot{y}} \right] \right) \dot{y} + \frac{\partial T}{\partial \dot{y}}\ddot{y} - \frac{dU}{dy}\dot{y} \\
&= \frac{d}{dt}\left[\frac{\partial T}{\partial \dot{y}} \right] \dot{y} + \frac{\partial T}{\partial \dot{y}}\ddot{y} \\
&= \frac{d}{dt}\left[\frac{\partial T}{\partial \dot{y}}\dot{y} \right].
\end{aligned}
$$

Hence, the object moves through time in such a way that

$$
\frac{d}{dt}\left(L - \frac{\partial T}{\partial \dot{y}}\dot{y} \right) = 0,
$$

or equivalently,

$$
L - \frac{\partial T}{\partial \dot{y}}\dot{y} = \text{constant} = -\mathcal{H}.
$$

The $\mathcal{H}$ in the last equality is the value of the object's *Hamiltonian*. The Hamiltonian is dependent on the object's dynamical conditions (kinetic energy and potential energy). Different objects may therefore have different values for the Hamiltonian, but the same object maintains a single value of the Hamiltonian throughout its motion.

We will now show that for the simple dynamics under consideration in this Module, the Hamiltonian $\mathcal{H}$ represents total energy (that is, it is the sum of the kinetic energy T and potential energy U). We assume that the kinetic energy of an object is given by $T = \frac{1}{2}m(\dot{y})^2$, and therefore

$$
\frac{\partial T}{\partial \dot{y}} = m\dot{y}.
$$

It follows that

$$
\dot{y}\frac{\partial T}{\partial \dot{y}} = m\dot{y}^2 = 2T.
$$

Hence,

$$
-\mathcal{H} = L - 2T = T - U - 2T.
$$

Our desired result follows by elementary algebra:

$$
\mathcal{H} = T + U.
$$

Since the value of $\mathcal{H}$ remains constant when describing the motion of an object, the total energy of the object must be conserved.

Whereas the Lagrangian is expressed as a function of y, $\dot{y}$, and t, the Hamiltonian should be expressed as a function of q, p and t, where

$$q = y, \qquad p = \frac{\partial L}{\partial \dot{y}}.$$

(Later, when defining Hamilton's equations, we shall see that the use of "conjugate variables" p and q gives rise to a nice symmetry in expressions for $\partial \mathcal{H}/\partial q$ and $\partial \mathcal{H}/\partial p$.)

In our case, $L = \frac{1}{2}m(\dot{y})^2 - mgy$, so that $p = \partial L/\partial \dot{y} = m\dot{y}$. Thus, the Hamiltonian may be expressed as

$$\mathcal{H} = \mathcal{H}(q, p, t) = \frac{p^2}{2m} + mgq.$$

(Here there is no direct dependence of the Hamiltonian on time t. In general, the Hamiltonian of an isolated dynamical system without energy dissipation is independent of time.)

This Hamiltonian arises in the idealized motion of ball with mass m falling freely with constant gravitational acceleration $-g$ (i.e. we neglect air resistance). For simplicity, we take $m = 1$ and $g = 1$. Let $y(t)$ be the height of a ball at time t. In Newtonian mechanics, by anti-differentiating the acceleration, we obtain both the height and the velocity of the ball at arbitrary time t given the initial velocity v_0 and the initial position y_0:

$$\frac{d^2 y}{dt^2} = -1, \qquad \frac{dy}{dt} = -t + v_0, \qquad y(t) = -\frac{1}{2}t^2 + v_0 t + y_0.$$

Thus, we may describe the motion of the ball as tracing the right half of a vertical parabola in which the height y of the ball is on the vertical axis and the time t is on the horizontal axis (see **Figure 1**).

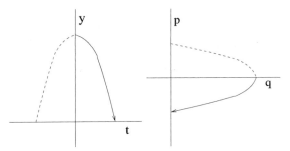

Figure 1. Freely-falling ball's motion described in terms of time t vs. height y (left) and in terms of height q vs. momentum p (right).

In Hamiltonian dynamics, we describe the motion of the ball using the coordinates $(q(t), p(t))$, where $q(t)$ is the position of the ball at time t (i.e., $q(t) = y(t)$) and $p(t)$ is, in this simple case, the momentum of the ball at time t

(i.e., $p(t) = m\dot{y} = \dot{y}$ since $m = 1$). The motion of the ball in the q-p plane traces the bottom half of a horizontal parabola (see **Figure 1** and **Exercise 3**). The q-p plane is called a *two-dimensional phase space*, and it is the setting in which we discuss Liouville's theorem in the next section.

Since the Hamiltonian $\mathcal{H}$ is the sum of the kinetic and potential energies ($\mathcal{H} = p^2/2 + q$), one can verify that (**Exercise 4**)

$$\frac{\partial \mathcal{H}}{\partial q} = -\frac{dp}{dt}, \qquad \frac{\partial \mathcal{H}}{\partial p} = \frac{dq}{dt}. \qquad \text{(Hamilton's equations)}.$$

These equations are fundamental in Hamiltonian dynamics; we use them in the next section when we derive Liouville's theorem.

Exercises

2. Consider a ball with mass m that falls freely under the influence of constant gravity (i.e., without air resistance). Let $y(t)$ ($t \geq 0$) be the height of the ball at time t. The kinetic energy of the ball is

$$T(\dot{y}) = \tfrac{1}{2}m(\dot{y})^2,$$

and its potential energy is

$$U(y) = mgy,$$

where $-g$ is the constant acceleration due to gravity.

a) Find a simple differential equation for y that is obtainable from Euler's equation.

b) Check that the the total energy $E = T + U$ is conserved, by computing dE/dt and simplifying using the answer to part a).

3. Consider a ball with unit mass that at $t = 0$ is dropped from rest at a height of 1, so that $(q_0, p_0) = (1, 0)$. Assuming constant acceleration $-g = -1$, show that in the two-dimensional q-p phase plane, the freely-falling ball's motion is along the horizontal parabola $q = 1 - \tfrac{1}{2}p^2$.

4. Verify Hamilton's equation for the case of a freely-falling ball. That is, show that

$$\frac{\partial \mathcal{H}}{\partial q} = -\frac{dp}{dt}, \qquad \frac{\partial \mathcal{H}}{\partial p} = \frac{dq}{dt}.$$

4. Derivation of Liouville's Theorem in Two Dimensions

Intuitively, Liouville's theorem in two dimensions says that

a large collection of phase points will always occupy the same amount of area, no matter how the shape of the area they occupy may change.

In other words, the density of a large collection of phase points does not change with time, even though the region in phase space formed by these phase points usually will change with time. (see **Figure 2**).

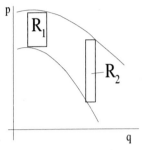

Figure 2. The phase points forming region R_1 at time t_1 form region R_2 at time t_2. The shape of the region changes but not the amount of area or the density of phase points.

We now introduce the concept of a *phase space density function* ρ, which may be evaluated at any phase point (q, p) and at any time t. For a small region in phase space, the phase density is the number of phase points in that region divided by the area of the region. Since the density ρ is dependent on q, p, and t, we write $\rho = \rho(q, p; t)$. Furthermore, we are interested in how the density changes as we move with the flow of points through phase space. In other words, we must compute the change in value of the density ρ as we move with a phase point along its path $\big(q(t), p(t)\big)$ through phase space. Since we allow q and p to change with time, we write $\rho = \rho\big(q(t), p(t); t\big)$.

Liouville's theorem in two dimensions asserts that the total derivative $d\rho/dt$ is zero. This means that the density will remain constant in the following sense. If at time t_0 we evaluate the density of an object at its phase point $\big(q(t_0), p(t_0)\big)$, and at any later time t we evaluate the density at the object's new phase point $\big(q(t), p(t)\big)$, we will obtain the same number.

By the chain rule for derivatives, we have

$$\frac{d\rho}{dt} = \frac{\partial\rho}{\partial q}\dot{q} + \frac{\partial\rho}{\partial p}\dot{p} + \frac{\partial\rho}{\partial t}.$$

Hence, one way of proving that $d\rho/dt$ equals zero is to show that

$$\frac{\partial\rho}{\partial t} = -\left(\frac{\partial\rho}{\partial q}\dot{q} + \frac{\partial\rho}{\partial p}\dot{p}\right),$$

where $\partial\rho/\partial t$ is the rate of change of phase density with time if we keep fixed the point of evaluation (q, p). To compute $\partial\rho/\partial t$, we consider the flow of phase points through a small rectangular area element of phase space. We position the bottom left corner of the rectangle at (q, p) and give the area element a horizontal length dq and vertical length dp (see **Figure 3**), so that the area of the rectangle is $dq\,dp$. Phase points flow in and out of this rectangle, with the net

7

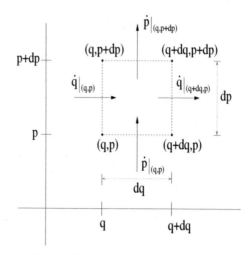

Figure 3. Diagram for computation of $\partial\rho/\partial t$.

rate of change (i.e., the increase in the number of phase points per unit time) for this area element being given by $\partial\rho/\partial t\ dq\ dp$.

The number of phase points flowing in through the bottom of the rectangle per unit time is given by the expression $\rho\dot{p}\ |_{(q,p)}\ dq$ (the notation $|_{(q,p)}$ indicates that both ρ and $\dot{p}$ are to be evaluated at the phase point (q,p)). This expression is the product of $\dot{p}$ (the vertical component of the flow velocity at the point (q,p)) with the horizontal length dq and the density function ρ. Similarly, the amount flowing in the left is $\rho\dot{q}\ dp$, where we keep in mind that both the density ρ and speed $\dot{q}$ are evaluated at the point (q,p).

Assuming that the vertical component of velocity of the phase points is constant along the top of the rectangle, the amount flowing out the top is $\rho\dot{p}\ |_{(q,p+dp)}\ dq$. Analogous to the fact that the linear approximation $f(x+h)\approx f(x)+f'(x)h$ becomes exact as h approaches zero, the linear approximation

$$\rho\dot{p}\ |_{(q,p+dp)}\approx \rho\dot{p}\ |_{(q,p)} +\frac{\partial}{\partial p}\ [\rho\dot{p}]\ |_{(q,p)}\ dp$$

becomes exact as dp approaches zero. It follows that the number of phase points flowing out the top is given by

$$\rho\dot{p}\ |_{(q,p)}\ dq + \frac{\partial}{\partial p}\ [\rho\dot{p}]\ |_{(q,p)}\ dp\ dq$$

where the point of evaluation is (q,p). The number of phase points out the right can be computed using the same reasoning and is therefore given by

$$\rho\dot{q}\ dp + \frac{\partial}{\partial q}\ [\rho\dot{q}]\ dq\ dp,$$

where the point of evaluation, (q,p), has been assumed.

The total change in number of phase points in the area element per unit time is equal to the rate phase points flow in minus the rate that they flow out. Hence, we obtain the following expression for $\partial\rho/\partial t\ dq\ dp$:

$$\frac{\partial\rho}{\partial t}\ dp\ dq = \rho\dot{p}\ dq + \rho\dot{q}\ dp - \left(\rho\dot{p}\ dq + \frac{\partial}{\partial p}\left[\rho\dot{p}\right]\ dp\ dq\right)$$

$$- \left(\rho\dot{q}\ dp + \frac{\partial}{\partial q}\left[\rho\dot{q}\right]\ dq\ dp\right).$$

The first term in the right side of the equation represents the amount that flows in the bottom, the second term represents the amount that flows in the left, the third represents the amount that flows out the top, and the last term represents the amount that flows out the right. Cancelling and simplifying leads to

$$\frac{\partial\rho}{\partial t}\ dp\ dq = -\left(\frac{\partial}{\partial p}\left[\rho\dot{p}\right] + \frac{\partial}{\partial q}\left[\rho\dot{q}\right]\right)\ dq\ dp,$$

$$\frac{\partial\rho}{\partial t} = -\left(\frac{\partial}{\partial p}\left[\rho\dot{p}\right] + \frac{\partial}{\partial q}\left[\rho\dot{q}\right]\right),$$

$$\frac{\partial\rho}{\partial t} = -\left(\frac{\partial\rho}{\partial p}\dot{p} + \rho\frac{\partial\dot{p}}{\partial p} + \frac{\partial\rho}{\partial q}\dot{q} + \rho\frac{\partial\dot{q}}{\partial q}\right).$$

The last step uses the product rule for partial derivatives.

Next, because $\dot{p} = -\partial\mathcal{H}/\partial q$ and $\dot{q} = \partial\mathcal{H}/\partial p$ in Hamiltonian dynamics, we may substitute, so that

$$\frac{\partial\rho}{\partial t} = -\left(\frac{\partial\rho}{\partial p}\dot{p} + \rho\frac{\partial}{\partial p}\left[-\frac{\partial\mathcal{H}}{\partial q}\right] + \frac{\partial\rho}{\partial q}\dot{q} + \rho\frac{\partial}{\partial q}\left[\frac{\partial\mathcal{H}}{\partial p}\right]\right).$$

The second and fourth terms cancel due to the equality of mixed partials, so we get

$$\frac{\partial\rho}{\partial t} = -\left(\frac{\partial\rho}{\partial p}\dot{p} + \frac{\partial\rho}{\partial q}\dot{q}\right).$$

Substituting this expression for $\partial\rho/\partial t$ into the total derivative $d\rho/dt$ given at the onset results in

$$\frac{d\rho}{dt} = \frac{\partial\rho}{\partial q}\dot{q} + \frac{\partial\rho}{\partial p}\dot{p} + \frac{\partial\rho}{\partial t}$$

$$= \frac{\partial\rho}{\partial q}\dot{q} + \frac{\partial\rho}{\partial p}\dot{p} - \left(\frac{\partial\rho}{\partial p}\dot{p} + \frac{\partial\rho}{\partial q}\dot{q}\right)$$

$$= 0.$$

This proves Liouville's theorem and confirms that the density does not change if at the initial time t_0 we evaluate it at the point $(q(t_0), p(t_0))$ and at any later time t we evaluate it at the point $(q(t), p(t))$.

There is a simple relationship between the density $\rho(q, p; t)$ and a probability distribution $f(q, p; t)$. Let N be the total number of phase points in the phase space. Then we have

$$f(q, p; t) = \frac{\rho(q, p; t)}{N}.$$

In other words, the number of phase points in an area $dq\, dp$ is given by $\rho\, dq\, dp$, whereas the probability of finding a phase point in the same area is $f\, dq\, dp$. It is then simple to show (**Exercise 5**) that Liouville's theorem may be expressed in its probability distribution form as $df/dt = 0$. The two-dimensional probability distribution surface $z = f(p, q; t)$ moves with the flow of phase points in such a way that the height of the surface above each particular phase point remains constant as that phase point moves through phase space.

Exercise

5. Prove the probability distribution form of Liouville's theorem. That is, show that $df/ft = 0$, where $f(p, q; t)$ is a probability distribution function for a two-dimensional phase space. (Josiah Willard Gibbs (1839–1903) of Yale University is given credit for being the first to derive explicitly the general probability distribution form of Liouville's theorem [Binney and Tremaine 1987]. Gibbs was also the first to recognize that this theorem could be applied to astrophysics. We discuss this idea further in **Section 6**.)

5. A Simple Illustration Using Three Freely-Falling Balls

Let $\mathbf{F}(q, p)$ be the vector field that gives the "velocity" of a phase point (q, p) in two-dimensional phase space. In other words,

$$\mathbf{F}(q, p) = \left\langle \frac{dq}{dt}, \frac{dp}{dt} \right\rangle.$$

Hence we may write

$$\mathbf{F}(q, p) = \left\langle \frac{\partial \mathcal{H}}{\partial p}, -\frac{\partial \mathcal{H}}{\partial q} \right\rangle.$$

The divergence of $\mathbf{F}$ is zero, as we can see from the following:

$$
\begin{aligned}
\nabla \cdot \mathbf{F} &= \frac{\partial}{\partial q}\left[\frac{\partial \mathcal{H}}{\partial p}\right] + \frac{\partial}{\partial p}\left[-\frac{\partial \mathcal{H}}{\partial q}\right] \\
&= \frac{\partial^2 \mathcal{H}}{\partial q\, \partial p} - \frac{\partial^2 \mathcal{H}}{\partial p\, \partial q} = 0.
\end{aligned}
$$

A vector field whose divergence is zero at each point is said to be *incompressible*. This means that a collection of phase points forming a region with a certain area

10

at time t_0, will move through phase space in such a way that the area occupied by these points will be the same at any later time t. This result may also be obtained as a special case of Liouville's theorem in which the distribution f is uniform (or constant) on an area region A in q-p phase space (**Exercise 7**).

We may illustrate the incompressibility of phase flow in the following way. Let us consider three freely-falling balls A, B, and C whose initial positions in phase space are

$$A_0 = (1, 0), \qquad B_0 = (1.5, -1), \qquad C_0 = (2.5, -1).$$

We may think of balls B and C as having been released from heights $y = 2$ and $y = 3$, respectively, one second before ball A is released from rest at height $y = 1$. Note that the area of the triangle $\triangle A_0 B_0 C_0$ is $(0.5)(1)(1) = 0.5$ square units (see **Figure 4**).

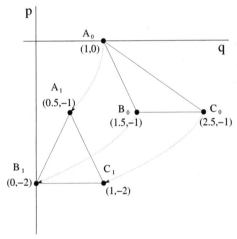

Figure 4. A special case of Liouville's theorem implies that the areas of triangles $A_0 B_0 C_0$ and $A_1 B_1 C_1$ must be the same.

A phase point initially at (q_0, p_0) will be at the new phase point

$$(q(t), p(t)) = \left(-\tfrac{1}{2}t^2 + p_0 t + q_0, \; p_0 - t\right)$$

after an amount of time t. Hence, after one second, the phase points that form $\triangle A_0 B_0 C_0$ will form $\triangle A_1 B_1 C_1$, where

$$
\begin{aligned}
A_1 &= (-0.5 + 1, -1) = (0.5, -1) \\
B_1 &= (-0.5 - 1 + 1.5, -2) = (0, -2) \\
C_1 &= (-0.5 + 2.5 - 1, -2) = (1, -2).
\end{aligned}
$$

With a little thought about the physical motion of freely-falling balls, one realizes that all of the representative phase points on or inside $\triangle A_0 B_0 C_0$ at time

11

$t = 0$ must be on or inside $\triangle A_1 B_1 C_1$ at time $t = 1$. Furthermore, no point outside $\triangle A_0 B_0 C_0$ at time $t = 0$ will be on or inside $\triangle A_1 B_1 C_1$ at time $t = 1$. Since the flow of phase points is incompressible ($\nabla \cdot \mathbf{F} = 0$), the areas of the two triangles must be the same. Indeed, both triangles have an area of 0.5 square units as can be seen in **Figure 4**.

Exercises

6. Find $\triangle A_2 B_2 C_2$, which is formed after two seconds by the points of the phase space initially in $\triangle A_0 B_0 C_0$, and show that the area of $\triangle A_0 B_0 C_0$ equals the area of $\triangle A_2 B_2 C_2$.

7. Let $A(t_0)$ be an area region formed by a collection of phase points at time t_0. Let $A(t)$ be the area region occupied by these phase points at any later time t. Use the probability distribution form of Liouville's theorem to prove that area($A(t_0)$) = area($A(t)$).

6.   Further Applications

6.1   Charged-Particle Accelerators

Particle accelerators attempt to create subatomic particles and interactions; in addition, they are used in a remarkable variety of practical applications (see [Wilson 2001]), including

- manufacturing of such diverse products as computer disks, shrink-wrap, automobile tires and telephone cables;

- purification of food stuffs, drinking water and surgical tools; and

- in medical procedures such as diagnostic imaging systems and radiation therapy techniques.

A simple case occurs when the beam circulates around an accelerator with constant energy. A cross section of such a beam may be represented by an ensemble of particles in two-dimensional q-p phase space, forming an elliptical region. Because the particles are charged, they can be focused by an arrangement of magnets, much as a light beam can be focused by geometric lenses. Since the magnetic force is conservative, by Liouville's theorem the shape of the elliptical-region in phase space may change during focusing but its area remains constant (**Figure 5**). An important quantity called the *emittance* of a beam is proportional to the area of the region occupied by the points in phase space. Hence, the emittance remains constant during the focusing of a constant energy beam.

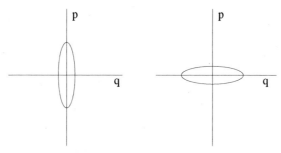

Figure 5. A cross section of particles in a constant-energy beam may be represented by an elliptic region in phase space. Liouville's theorem says that the area of the region occupied in phase space remains constant. Hence, the beam's emittance, which is proportional to the area, is also constant.

6.2   Galactic Dynamics

Two years after Liouville's death in 1882, Josiah W. Gibbs (1839–1903), a mathematical physicist at Yale University, recognized that the probability distribution form of Liouville's theorem could be applied to astronomy. Liouville's theorem arises within galactic dynamics in at least three different settings [Binney and Tremaine 1987]:

6.2.1   Analyzing Motion of Stars

A star moving within a galaxy may be represented by a phase point moving within a phase space with three position and three momentum coordinates. In this context, the number density function $\rho = \rho(q(t), p(t), t)$ with $q(t) = \big(q_1(t), q_2(t), q_3(t)\big)$ and $p(t) = \big(p_1(t), p_2(t), p_3(t)\big)$ is used to specify the number $\rho\, dq\, dp$ of stars within a small volume dq and momentum range dp centered at (q, p). Liouville's theorem asserts that $d\rho/dt = 0$, or, in other words, the number density remains the same around the phase point representing a given star.

6.2.2   Describing Macroscopic Properties

A galactic system with N stars may also be represented by a $6N$-dimensional phase space called a Γ-*space*; a point $(w_1, w_2, ..., w_N)$ in this phase space is called a Γ-*point*. The ith coordinate w_i of a Γ-point describes the position and momentum of a particular star. Different Γ-points may correspond to the same set of macroscopic galactic properties (density distribution, velocity distribution, number of binary stars etc.) Collectively, Γ-points giving rise to the same set of macroscopic properties are called an *ensemble*. The *N-particle distribution function* $f^{(N)} = f^{(N)}(w_1, ..., w_N, t)$ is then used to obtain the probability that a Γ-point belonging to an ensemble is found in a region D of Γ-space at time t. (One finds the probability by integrating $f^{(N)}$ over D.) The evolution of a galactic ensemble is therefore described by the evolution of its distribution $f^{(N)}$, the latter being constrained by the probability distribution form of Liouville's theorem $(df^{(N)}/dt = 0)$.

6.2.3   Modeling the Dynamics in a Cluster of Galaxies

On a massive scale, the dynamical motion of thousands of galaxies comprising a gigantic cluster may be modeled in a similar way to the motion of stars within a galaxy, hence creating a third setting for Liouville's theorem to be applied. Clusters continue to be an important site for the investigation into the existence, nature and distribution of enigmatic dark matter. Thirteen different clusters, including the Coma cluster and Abel 2142, have been studied by the Chandra X-ray observatory [Harvard-Smithsonian Center for Astrophysics 2003] since it was first deployed in 1999 by the space shuttle Columbia.

7.   Solutions to the Exercises

1. Euler's equation implies that $\ddot{y} = 1$. It follows by antidifferentiation that $y = y_0 + \dot{y}_0 t + \frac{1}{2}t^2$.

2. **a)** According to Hamilton's principle, the ball falls so that the integral of the difference between its kinetic and potential energy is minimized. Letting $f(y, \dot{y}; t) = \frac{1}{2}m\dot{y}^2 - mgy$, Euler's condition becomes

$$-mg - \frac{d}{dt}(m\dot{y}) = 0,$$
$$mg + m\ddot{y} = 0,$$
$$\ddot{y} = -g.$$

   b) $E = \frac{1}{2}m\dot{y}^2 + mgy$, so $dE/dt = m\dot{y}\ddot{y} + mg\dot{y} = m\dot{y}(-g) + mg\dot{y} = 0$.

3. The motion in the q-p phase plane will be along the parabola $q = -\frac{1}{2}p^2 + 1$, since

$$\frac{dy}{dt} = p = -t,$$
$$y = q = -\frac{1}{2}t^2 + 1 = -\frac{1}{2}p^2 + 1.$$

4. Assume that the ball has mass m and constant acceleration $-g$. The Hamiltonian $\mathcal{H}$ is given by $\mathcal{H} = p^2/2m + mgq$. Hence, $\partial\mathcal{H}/\partial q = mg$ and $\partial\mathcal{H}/\partial p = p/m$. We also have that

$$\text{Force } = -mg = \frac{dp}{dt},$$
$$\text{Velocity } = \frac{dq}{dt} = \frac{p}{m}.$$

   It follows that $\partial\mathcal{H}/\partial q = -dp/dt$ and $\partial\mathcal{H}/\partial p = dq/dt$.

5.
$$\frac{df}{dt} = \frac{1}{N}\frac{d\rho}{dt} = 0.$$

6. $A_2 = (-1, -2)$, $B_2 = (-2.5, -3)$, $C_2 = (-1.5, -3)$. Area of $\triangle A_2 B_2 C_2$ is 0.5.

7. Let area$(A(t_0)) = k$ and let $f(p, q, t_0) = 1/k$ for all $(p, q) \in A(t_0)$ and zero otherwise. By Liouville's theorem, $f(p, q, t) = 1/k$ for all $(p, q) \in A(t)$ and zero otherwise. Since the total probability is equal to one, area $A(t)) = k$, and we conclude that area$(A(t)) = $ area$(A(t_0))$.

References

Binney, James and Scott Tremaine. *Galactic Dynamics*. Princeton, NJ: Princeton University Press 1987.

Harvard-Smithsonian Center for Astrophysics. 2003. Chandra Images by Category. http://chandra.harvard.edu/photo/category/galaxyclusters. html .

Marion, Jerry B., and Stephen Thornton. 1995. *Classical Dynamics of Particles and Systems*. 4th ed. San Diego, CA: Harcourt Brace Jovanovich.

Wilson, E. 2001. *An Introduction to Particle Accelerators*. New York: Oxford University Press.

Acknowledgments

The authors thank Dr. Akira Isihara, a distinguished theoretical physicist in statistical mechanics, for inspiring the writing of this Module, and a referee for helpful corrections.

About the Authors

Paul Isihara, Professor of Mathematics at Wheaton College (IL), has a special interest in doing expository and creative research projects with undergraduates.

Nat Stapleton (mathematics '04) and Benjamin Noonan (physics '04) worked on this Module while participants in the 2003 Wheaton College Summer Science Research Program.

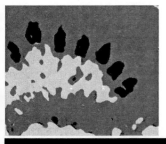

AUTHORS:
Marie Vanisko
(Mathematics, Engineering,
Physics, & Computer
Science)
Carroll College
Helena, MT 59625
mvanisko@carroll.edu

EDITORS:
Chris Arney,
Kathleen Snook,
and Steve Horton
Dept. of Mathematical
Sciences
U.S. Military Academy
West Point, NY

CONTENTS

Who Falls Through the Healthcare Safety Net?

MATHEMATICS CLASSIFICATIONS:
This project is appropriate for a mathematics course intended to serve students majoring in fine arts and the humanities, frequently referred to as a "liberal arts" mathematics course.

DISCIPLINARY CLASSIFICATIONS:
Health Information Management, Business, Ethics, Sociology, Political Science

PREREQUISITE SKILLS:
An understanding of basic statistical terms and charts

PHYSICAL CONCEPTS EXAMINED:
Actual data sets from the Census Bureau are analyzed concerning those without health insurance, first in general, then relative to those in poverty, and finally broken down by state. The emphasis is on interpreting the statistics and explaining them to others.

COMPUTING REQUIREMENT:
Either a graphing calculator or a computer with spreadsheet, computer algebra system, and/or statistical package

UMAP/ILAP Modules 2002–2003: Tools for Teaching, 69–76. Reprinted from *The UMAP Journal* 23 (1) (2002) 75–82.

Contents
1. Introduction
2. Instructions
3. Requirements
Instructors' Comments and Solutions
About the Author

1.   Introduction

Many of us take our health insurance for granted; but according to the U.S. Census Bureau report for 1999, more than 15% of the people in the United States have no health insurance, public or private. That is more than 42 million people, many of whom are children. With increasing medical costs, those without insurance frequently have to forego medical treatment or must give up their life savings to pay for it.

This Interdisciplinary Lively Application Project (ILAP) explores groups of people without health insurance in an effort to analyze the problem. All data are from the U.S. Census Bureau.

2.   Instructions

The group must provide detailed written reports, as instructed in the requirements, and be prepared to discuss the results in class.

This ILAP, with periodically updated data, is at http://web.carroll.edu/mvanisko/default.htm. Select Carroll ILAPs on the menu page and go to the ILAPs under Distribution of Wealth. The tables can be copied and pasted into an Excel spreadsheet. If direct pasting does not put the data into columns:

- Select (highlight) the first column (already highlighted after pasting) and deselect the other columns.

- Under the Datamenu item, select Text to Columns (the default is Fixed and that frequently works, but you can check and see).

- Click on Finish—and all the data are in columns. There may be a couple of details that you have to clean up, but they are minor.

3.   Requirements

Requirement 1

Table 1 gives a breakdown of individuals not covered by health insurance in 1999. Develop a profile for those without insurance, show the results graphically if appropriate, and provide a rationale for why the results turned out as they did.

Requirement 2

Table 2 gives a breakdown of 1999 health insurance coverage status for persons living below the poverty level by selected characteristics. This group represents 11.8% of the United States population, and some people in this group do not qualify for Medicaid (government health insurance for the poor) or Medicare (government health insurance for the elderly). Develop a profile for those without insurance in this group. Discuss why the results turned out as they did. Be sure to look at the percentage who are under 18. Compare and contrast these results to those in **Requirement 1**. How is the profile different for those in poverty than for the general population?

Requirement 3

Go to the Internet site http://www.census.gov and examine other Census Bureau data concerning poverty. Determine how "poor" is defined by the Census Bureau. Write a two- to three-page paper on what you have found and be prepared to present your results to the class.

Requirement 4

After completing the first three requirements, summarize your results in the form of a newspaper article that is approximately two pages in length. The goal is to see how effective you can be in translating the numbers into meaningful rhetoric. After all the reports are in, you will select one group's report to submit to the school and/or the city newspaper.

Requirement 5

Table 3 provides a breakdown regarding those without health insurance in each state. Categorize each state in terms of its location. Use **Table 3** to construct boxplots and scatterplots to explore differences in statistics among the states relative to their location and to their population. Compute the mean and median for the group as a whole and for each region of the country and compare the measures of central tendency within and across regions. Provide plausible explanations for variations you see.

Requirement 6

How could the information given about health insurance coverage be used to convince someone that a change in national health policy is needed? How could the same information be used to convince someone that a change is not needed? What does this tell you about this type of information?

Table 1.

All Persons Not Covered by Health Insurance, by Selected Characteristics: 1999.

Source: U.S. Census Bureau, Current Population Surveys, March 1999 and 2000.

Numbers in thousands

Characteristic	Total Number	Not covered by health insurance Number Not Covered	Percent Not Covered
All	274,087	42,554	15.5
Sex			
Male	133,933	22,073	16.5
Female	140,154	20,481	14.6
Age			
Under 18 years	72,325	10,023	13.9
18 to 24 years	26,532	7,688	29.0
25 to 34 years	37,786	8,755	23.2
35 to 44 years	44,805	7,377	16.5
45 to 64 years	60,018	8,288	13.8
65 years and over	32,621	422	1.3
Race and Hispanic Origin			
White	224,806	31,863	14.2
White, not of Hisp. Origin	193,633	21,363	11.0
Black	35,509	7,536	21.2
Asian / Pacific Islander	10,925	2,272	20.8
Hispanic origin[1]	32,804	10,951	33.4
Education (persons aged 18 and over)			
No high school diploma	34,087	9,111	26.7
High school graduate only	66,141	11,619	17.6
Some college, no degree	39,940	6,051	15.2
Associate degree	14,715	1,902	12.9
Bachelor's degree or higher	46,880	3,848	8.2
Work Experience (persons aged 18 to 64)			
Worked during year	139,218	24,187	17.4
Worked full-time	115,973	18,984	16.4
Worked part-time	23,245	5,204	22.4
Did not work	29,923	7,921	26.5
Nativity			
Native	245,708	33,089	13.5
Foreign-born	28,379	9,465	33.4
Naturalized citizen	10,622	1,900	17.9
Not a citizen	17,758	7,565	42.6
Household Income			
Less than $25,000	64,628	15,577	24.1
$25,000–$49,999	77,119	13,996	18.2
$50,000–$74,999	56,873	6,706	11.8
$75,000 or more	75,467	6,275	8.3

[1]Persons of Hispanic origin may be of any race.

Table 2.

Poor Persons Not Covered by Health Insurance, by Selected Characteristics: 1999.

Source: U.S. Census Bureau, Current Population Surveys, March 1999 and 2000.

Numbers in thousands

Characteristic	Total Number	Not covered by health insurance Number Not Covered	Percent Not Covered
All	32,258	10,436	32.4
Sex			
Male	13,813	4,830	35.0
Female	18,445	5,606	30.4
Age			
Under 18 years	12,109	2,825	23.3
18 to 24 years	4,603	2,088	45.4
25 to 34 years	3,968	2,059	51.9
35 to 44 years	3,733	1,672	44.8
45 to 64 years	4,678	1,686	36.0
65 years and over	3,167	107	3.4
Race and Hispanic Origin			
White	21,922	7,271	33.2
White, not of Hisp. Origin	14,875	4,158	28.0
Black	8,360	2,347	28.1
Asian / Pacific Islander	1,163	485	41.7
Hispanic origin[1]	7,439	3,254	43.7
Education (persons aged 18 and over)			
No high school diploma	7,888	2,876	36.5
High school graduate only	6,810	2,611	38.3
Some college, no degree	3,162	1,278	40.4
Associate degree	836	324	38.8
Bachelor's degree or higher	1,452	521	35.9
Work Experience (persons aged 18 to 64)			
Worked during year	8,649	4,104	47.5
Worked full-time	5,582	2,654	47.5
Worked part-time	3,066	1,450	47.3
Did not work	8,333	3,400	40.8
Nativity			
Native	27,507	7,817	28.4
Foreign-born	4,751	2,619	55.1
Naturalized citizen	968	347	35.9
Not a citizen	3,783	2,271	6.0

[1] Persons of Hispanic origin may be of any race.

Table 3.

Number of Persons Covered and Not Covered by Health Insurance by State in 1999

Source: U.S. Census Bureau, Current Population Surveys, March 1999 and 2000.

State	Total thousands	Covered thousands	Not Covered thousands	Not Covered percent
United States	272,691	230,424	42,267	15.5
Alabama	4,370	3,745	625	14.3
Alaska	620	501	118	19.1
Arizona	4,778	3,765	1,013	21.2
Arkansas	2,551	2,176	375	14.7
California	33,145	26,417	6,728	20.3
Colorado	4,056	3,375	681	16.8
Connecticut	3,282	2,960	322	9.8
Delaware	754	668	86	11.4
District of Columbia	519	439	80	15.4
Florida	15,111	12,210	2,901	19.2
Georgia	7,788	6,534	1,254	16.1
Hawaii	1,185	1,054	132	11.1
Idaho	1,252	1,013	239	19.1
Illinois	12,128	10,418	1,710	14.1
Indiana	5,943	5,301	642	10.8
Iowa	2,869	2,631	238	8.3
Kansas	2,654	2,333	321	12.1
Kentucky	3,961	3,387	574	14.5
Louisiana	4,372	3,388	984	22.5
Maine	1,253	1,104	149	11.9
Maryland	5,172	4,561	610	11.8
Massachusetts	6,175	5,527	648	10.5
Michigan	9,864	8,759	1,105	11.2
Minnesota	4,776	4,393	382	8.0
Mississippi	2,769	2,309	460	16.6
Missouri	5,468	4,998	470	8.6
Montana	883	719	164	18.6
Nebraska	1,666	1,486	180	10.8
Nevada	1,809	1,435	375	20.7
New Hampshire	1,201	1,079	123	10.2
New Jersey	8,143	7,052	1,091	13.4
New Mexico	1,740	1,291	449	25.8
New York	18,197	15,212	2,984	16.4
North Carolina	7,651	6,473	1,178	15.4
North Dakota	634	559	75	11.8
Ohio	11,257	10,018	1,238	11.0
Oklahoma	3,358	2,770	588	17.5
Oregon	3,316	2,832	484	14.6
Pennsylvania	11,994	10,867	1,127	9.4
Rhode Island	991	922	68	6.9
South Carolina	3,886	3,202	684	17.6
South Dakota	733	647	87	733
Tennessee	5,484	4,853	631	11.8
Texas	20,044	15,374	4,670	11.5
Utah	2,130	1,827	302	14.2
Vermont	594	521	73	12.3
Virginia	6,873	5,904	969	14.1
Washington	5,756	4,847	910	15.8
West Virginia	1,807	1,498	309	17.1
Wisconsin	5,250	4,673	578	11.0
Wyoming	480	402	77	16.1

Title: **Who Falls Through the Healthcare Safety Net?**

Instructors' Comments and Solutions

This ILAP requires minimal computations. The focus here is to raise the mathematical literacy of the intended audience, so that they might become a more "informed citizenry" (in the words of Thomas Jefferson). A "liberal arts" mathematics course is generally intended for students in English, fine arts, history, etc.; such individuals generally avoid mathematics and having to sift through numbers to see what the story is behind the numbers.

This ILAP provides current information about the state of the nation regarding those without health insurance. The focus of the first four Requirements is summed up in **Requirement 4**. Students must study the data and write intelligently about where the problem might be. In a similar way, **Requirement 6** asks students to interpret the data pertaining to individual states.

Requirement 5 has formal mathematical content and suggested solutions follow in **Table S1** and **Figure S1**.

Table S1.

State summary statistics.

Statistic	% Uninsured
Mean	14.4
Median	14.2
Std. Deviation	4.3
Minimum	6.9
Maximum	25.8
Percentiles:	
25th	11.1
50th	14.2
75th	17.1

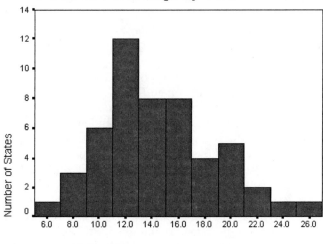

Figure S1. Histogram of state percentages.

About the Author

Marie Vanisko is a professor of mathematics at Carroll College in Helena, Montana, where she has taught for more than 30 years. As a co-director of the NSF Project INTERMATH at Carroll, she has been a primary mover in initiating the writing of interdisciplinary projects (ILAPs), and she has taken a lead role in instituting curricular reform in the undergraduate mathematics program. Marie is co-author of the technology supplement that accompanies the 10th edition of Thomas's *Calculus* (2001) and has served as a judge in COMAP's Mathematical Contests in Modeling at both the undergraduate and the high school levels. In Spring 2002, she was a visiting professor at the Department of Mathematical Sciences at the U.S. Military Academy at West Point.

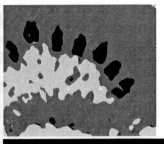

AUTHORS:
Lisa Pike
(Biology)

William P. Fox
(Mathematics)
bfox@fmarion.edu

Francis Marion University
Florence, SC

EDITOR:
Richard D. West
Francis Marion University
Florence, SC

CONTENTS

Stocking a Fish Pond

MATHEMATICS CLASSIFICATIONS:
Algebra, Statistics

DISCIPLINARY CLASSIFICATIONS:
Population Biology

PREREQUISITE SKILLS:
1. Algebra skills
2. Solving equations
3. Basic statistics

PHYSICAL CONCEPTS EXAMINED:
1. Biological populations
2. Mark/recapture technique

COMPUTING REQUIREMENT:
Graphing calculator

UMAP/ILAP Modules 2002–2003: Tools for Teaching, 77–90. Reprinted from *The UMAP Journal* 23 (2) (2002) 135–148. © Copyright 2002, 2003 by COMAP, Inc. All rights reserved.

Contents

Setting the Scene

A number of basic measurements are used in describing populations and populated communities. Among these are population density, abundance of particular species, distribution of species, population size, and population age structures. Ecologists call a total count of all humans in a population a *census*, but it is seldom possible to count every individual within a population (for example, there is debate about the accuracy of U.S. Census data.) At best, ecologists can look at a small portion of the population and make inferences about the whole. Environmental scientists use data like these as a baseline, for comparison to data taken after an environmental impact.

For example, there is debate about the disappearance of frog species throughout the world. Many populations are decreasing in number of individuals, and some species are going extinct, such as the golden toad of Costa Rica (*Bufo periglenes*). Is the decline in amphibian numbers due to habitat loss, global warming, pollution or ozone loss? And how do we know the population has declined or that the decline is significant?

How can we answer these questions about population change? One method involves capturing and marking individuals of a population. Marking individuals that you have caught is necessary if you are to replace the individuals back into the ecosystem in order to get population data. By marking individuals of a population and recording specific data, biologists may be able to infer information about age, longevity, growth rates, dispersal and home range of the population.

How are these individuals marked? Marking techniques may include the use of paints or dyes, tags of metal strapped around an animals' leg (birds), attached to a dorsal fin (fish), or tags that are inserted into a body cavity (snakes, fish) and detected using a radio-transmitter. Some animals are marked by clipping a toe (frogs, salamanders, lizards, and small mammals), which involves the removal of the distal part of one or more toes. Snakes can be marked by removing certain patches of scales.

More secretive animals can be marked using radioactive tracers placed in their food source. These tracers work well in tracking animals that have radically different phases in their lifestyles. For example, an adult butterfly can be fed radioactive material. This is incorporated into the egg, which then turns into a radioactive larvae and later into an adult.

A population estimate of a highly mobile species is usually done by the *mark/recapture technique* (also called the *Lincoln-Peterson method*). This involves capturing as many individuals from a population as possible, marking them as described above, and then releasing them. After a period of time, we again try to capture as many individuals from the population as possible. Then compare the number of individuals that were caught the second time (the ones that have the "mark") to the number of individuals that were only caught once (the ones that are unmarked). This technique can be used to record declines in amphibian (or other) populations and can also be used in a variety of other ways. For example, this technique can be used to support or allow bag limits or quotas to be set for hunting seasons. How many deer in a population can be culled without affecting the ability of the population to be sustainable, that is, to have enough deer to survive and be healthy year after year?

No technique for population estimation is foolproof, and many are biased (either underestimating or overestimating population sizes). So we generally want to calculate a confidence interval to estimate the population size. How sure are we that our estimate is correct? Estimates are generally more accurate with larger populations.

In-Class Motivating Example/Experiment

Part 1. The Experiment

Classroom Supplies Needed

A large bowl or jar and either

- Goldfish-brand snack food (white cheddar and regular flavors) or

- a bag of white beans and a bag of brown beans.

In the directions below, we assume that you are using the beans.

Procedure

1. Pour all the white beans into a bowl and ask, "How many are in the bowl?"

2. Sample lots of guesses from the students.

3. Take the average of their guesses and place on the board for later comparison.

4. Now, let's improve on our guesses. Ask a student to grab a large handful of white beans from the bowl and count that number.

5. Then have another student count out the same number of brown beans (as the handful of white beans) and replace the white beans removed from the bowl with the brown beans. Mix thoroughly.

6. Grab another large handful of beans from the bowl. Count the number of white and brown beans in that new handful.

7. Set up the following ratio:

$$\frac{\text{number of brown beans}}{\text{number of white beans}} = \frac{\text{number in first handful}}{\text{total}}.$$

8. Solve for the total:

$$\text{total} = \text{number in first handful} \times \frac{\text{number of white beans}}{\text{number of brown beans}}.$$

Note: This is an estimate. Repeat this experiment with a few students and keep the data for further comparison in later parts of this project. This exercise demonstrates an example of the mark/recapture technique as applied on a population of either Goldfish crackers or beans.

In the Biology Classroom

This exercise demonstrates an example of the mark/recapture technique used on meal worms (*Tenebrio*). *Tenebrio* are beetles that lay eggs, which hatch after 7 to 10 days. The young beetle, the larva, is the grub-like "meal worm," which will pupate after 2 to 3 weeks and metamorphose into the adult beetle.

1. Divide up all of the meal worms into cups, each lab group receiving a cup of worms. Count all your worms, and then mark them with a paint pen. Each group should use a different color. Let the paint dry.

2. Place all worms from all groups back into the large cup on the front desk, where your instructor will mix them up.

3. "Recapture" some meal worms. Your instructor will once again divide all the meal worms among the groups.

4. Count the worms again, and then count how many have "your" mark (your color of paint).

5. Then calculate a population size estimate using the following ratios:

$$\frac{M}{R} = \frac{t}{N}.$$

This reduces nicely to the following formula (and is the same ratio as with the mathematics classroom experiment):

$$N = \frac{Mt}{R},$$

where

N = population size estimate,

M = number marked when first caught,

t = total number caught in second sampling, and

R = number caught the second time that had your mark (from first capture).

6. Find this point estimate from your collected data.

In either experiment, the result is known as a *point estimate*. A point estimate is a single value that represents the most plausible value for a parameter of the population.

For either class experiment, here are some questions and additional work for the students.

- Have the students list any problems with using this point estimate as their guess.

- What assumptions did they make that allowed them to find this point estimate in the first place?

- Have the students collect data in this experiment and repeat the experiment. Then have them find the following descriptive statistics for their data:

  - mean

  - variance

  - standard deviation

Situation

Our neighbor, Mr. Bait, owns a large pond that he keeps stocked with trout. He sells fishing passes to tourists, and everyone seems to enjoy fishing at his pond. Mr. Bait needs to know if the fish are reproducing fast enough. Can he skip some years in restocking the pond, and only restock when necessary? Doing so would enable Mr. Bait to save on the costs of restocking and help increase his overall profits.

Mr. Bait hired an environmental team to collect some data and has hired your team to analyze the data. He needs your team to make predictions about the number of trout in his pond. The environmental team came out this past fall and captured 100 trout and marked each of them in the dorsal fin with a small metal tag. They returned in the spring and collected the data displayed in **Table 1**. Each day for six days an experiment was conducted. After each experiment, all trout were returned to the pond. No new trout were marked during this phase of the experiment. No fishing was allowed during the experiment and no restocking was done.

Table 1.

Mark/recapture (trout) data sheet, 100 individuals marked.

Day	Unmarked	Marked
1	790	10
2	710	20
3	630	20
4	810	50
5	610	60
6	630	10
Total:	4180	170

Requirements

Part 1

1. Determine the point estimate for the total number of trout in the pond from each experiment in **Table 1**.

2. Determine the mean, variance, and standard deviation from these point estimates for the total.

Part 2

In Part 1, your team has the data for the repeated mark/recapture experiment. You now have a mean, variance, and standard deviation for each data set. Let's use this information to produce a simulated result.

1. Build a simulation (using the means and standard deviations of your data) to provide guesses to the population. Do this simulation for 10, 100, and 1000 runs.

2. Plot the computer guesses (as a dotplot, a histogram, or a stem-and-leaf plot) from each set of runs (10, 100, and 1000) and then comment about the shape of the distribution for each run.

3. Is the shape symmetric? Is it approximately a bell-shaped curve? For which number of runs?

   The Central Limit Theorem states that if you sample either the average or the total of a population, then the distribution of the sample is approximately a normal distribution with mean and standard deviations as shown in **Table 2**.

4. Determine which of the simulated results from the runs of $n = 10$, 100, or 1000 best matches our formulas above. Explain why.

Table 2.

Parameters of Central Limit Theorem.

Population	Mean	Standard deviation
Mean	$\bar{x}$	$\dfrac{s}{\sqrt{n}}$
Total	$n\bar{x}$	$s\sqrt{n}$

Part 3

Your team needs to improve on these estimates. Almost any parameter that we might want to estimate has, as its set of possible values, an entire interval of numbers. Thus, a point estimate can be vastly improved through the use of a *confidence interval*. A confidence interval is an interval built around the parameter of interest and based upon sound statistical assumptions. Basic properties of confidence intervals for a population about a parameter are that:

- the population's distribution is approximately normal for the total or the average, and

- the value of the population standard deviation is known or can be found.

Your team needs to consider both 95% and 99% confidence intervals for the population total that will provide the following range:

Lower Confidence Limit (LCL) $\leq N \leq$ Upper Confidence Limit (UCL).

The mark/recapture analysis uses a "raw" formula to find the standard error (SE):

$$\text{SE} = \sqrt{\frac{Mt^2(t-R)}{R^3}}.$$

We can use this SE in a 95% confidence interval based upon 2 standard deviations using the following interval formula:

$$(N - 2\text{SE}, N + 2\text{SE}).$$

Assume that the data in **Table 3** are from a previous experiment for your team to practice with.

Table 3.

Sample population data.

M	t	R	SE	LCL	N	UCL
28	36	10	27	47	101	155

Using the data in the table, provide the following information:

1. What is your total population estimate?

2. Is it close to the actual population size?

3. Did you underestimate or overestimate?

4. What is the standard error?

5. What is the 95% confidence interval?

6. What is the 99% confidence interval?

7. What is the range for your population estimate?

8. What can you conclude about this data?

Repeat questions 1–8 for the total population data in Part 1.

Part 4

Further analysis of confidence intervals is required. Let's return to the idea of simulating experiments.

1. Based upon the knowledge of the Central Limit Theorem and the representative data in the table provided, simulate one hundred 95% confidence intervals from an approximately normal distribution with mean of 101 and SE of 27.

2. Plot each of these 100 confidence intervals. Count the confidence intervals that do not include the true value of 101. Does this number hold any significance? Since the estimate could change from week to week, depending on the proportion of marked individuals in the second sample, we could express total population size as a range of numbers with 95% degree of certainty. We can calculate the standard error and confidence range of the data using our equations:

3. For $N = 150$ and SE = 16, compute the 95% confidence interval and interpret the statistical meaning of that result.

4. Compute the 95% confidence interval for Mr. Bait's data and interpret this for Mr. Bait.

Part 5

Mr. Bait determines from historical data that if he has over 5,000 trout in his pond at the end of the year, then he should not have to restock for the next year. How often does he need to restock, and how many trout should he order to be stocked?

Part 6

Find the 90% confidence interval for the total population of bass from the data reported in a smaller pond owned by Mr. Bait's son. Support your interval estimate.

Table 4.

Mark/recapture (bass) data sheet, 50 individuals marked.

Day	Unmarked	Marked
1	0	0
2	90	2
3	40	2
4	190	10
5	100	5
6	10	0
Total:	430	19

Sample Solution

Part 1

1. There are 100 marked trout in the pond. Therefore, for the first estimate, we
have
$$\frac{100}{T} = \frac{10}{790},$$
so $T = 79000/10 = 7900$ trout.

This experiment was repeated 5 additional times with the rounded results
in **Table 5.**

Table 5.

Results of repetitions of the experiment.

Experiment	Unmarked	Marked	Original	Total
1	790	10	100	7900
2	710	20	100	3550
3	630	20	100	3150
4	810	50	100	1620
5	610	60	100	1017
6	630	10	100	6300

We round as estimates for the total number of trout the following values:
7900, 3550, 3150, 1620, 1017, and 6300.

2. The descriptive statistics, using EXCEL, on the total of these six experiments
are in **Table 6.**

Table 6.

Summary statistics.

Mean	3922.778
Standard Error	1094.743
Median	3350
Mode	N/A
Standard Deviation	2681.563
Sample Variance	7190780
Kurtosis	−1.07891
Skewness	0.607997
Range	6883.333
Minimum	1016.667
Maximum	7900
Sum	23536.67
Count	6
Confidence Level(95.0%)	2814.123

A *point estimate* a single point that represents the most plausible value that
describes an entire population, such as a mean, a total, or a standard error.

Part 2

Figures 1–3 show the graphs and estimates of the total for simulations for 10, 100, and 1000 runs using population size software.

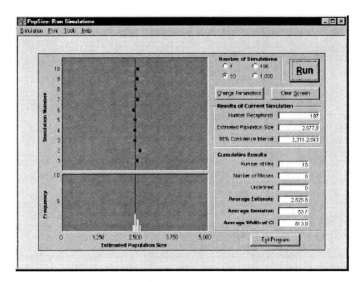

Figure 1. Results for 10 runs: Does not appear bell-shaped.

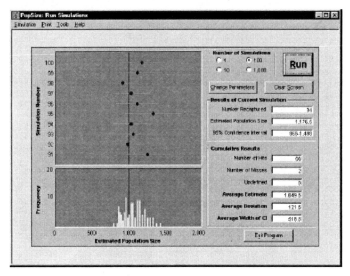

Figure 2. Results for 100 runs: Appears to be moving towards a bell-shaped curve but still very ragged.

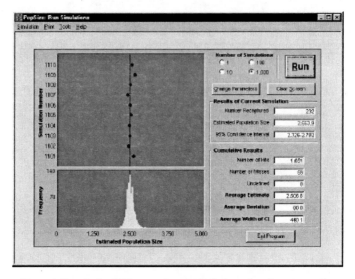

Figure 3. Results for 1000 runs: Appears bell-shaped and very smooth.

The frequency plots for the larger number of runs tend to support a normal distribution (bell-shaped curve). The 1000 runs is more bell-shaped than the 10 or 100 runs. Usually, it takes more than 30 runs for the average or the total to appear to be approximately a normal distribution. You can consult various probability and statistics texts to obtain different authors' views.

Part 3

In Part 2, we found that using the larger samples and invoking the Central Limit Theorem (CLT), the distribution of the sample mean appears to be normally distributed. We can use this information to find the confidence intervals:

- Using typical CI formulas and assuming the total is approximately normally distributed: For a 95% CI, the Z-value is 1.96.

$$\bar{x} \pm Z_{\alpha/2} = 3923 \pm 1.96(2681.49) = 3923 \pm 5255.72.$$

- Using the one sample method:

$$N \pm 2SE = 3923 \pm 2(1094) = 3923 \pm 2188 = [1735, 6111].$$

In comparison, we can do a better job in bounding the estimate by using more samples and assuming the normal distribution. For the data provided:

1. Population estimate: 100.8, or 101.

2. It is very close. Off by 0.2 prior to rounding.

3. Underestimated (slightly)

4. SE = 27.089

5. 95% CI is approximately $100.8 \pm 2(27.089) = 100.8 \pm 54.18 = [46.62, 154.98]$.

6. 99% CI is approximately $100.8 \pm 3(27.089)$.

7. Range: by the intervals, it is 47 to 155 (rounded).

8. We conclude that the data consists of a wide range of points and that the true average total might lie within this set.

Part 4

If we run one hundred 95% confidence intervals, we would expect that on average 95 of the 100 would contain the true population value. When you only obtain one CI, the true population value may either be in that interval or not; you do not know which.

Part 5

The estimate for the mean number of trout suggests that it is most likely below 5,000. It is suggested to restock about 1,100 trout next season and rerun this experiment.

Part 6

For the bass:
$$\frac{50}{T} = \frac{19}{43},$$
so $T = 113.2$. For a 90% CI, the Z-value is 1.65; the interval is $113.2 \pm 1.65(19.39)$.

About the Authors

Lisa Pike is an assistant professor of biology and has been teaching at Francis Marion University (FMU) for 10 years. She teaches Freshmen Biology, Environmental Science, and Marine Biology. Ms. Pike is also the FMU liaison for the Sustainable Universities Initiative and the faculty advisor for Ecology Club. She was born in Rochester NY; she received a B.S. In Biology from Binghamton University in 1989 and an M.S. In Marine Biology in 1991 from the University of North Carolina at Wilmington. After completing her master's degree, Ms. Pike was selected as a Rotary Scholar and spent a year at James Cook University of North Queensland, Australia.

Dr. William P. Fox received his B.S. degree from the United States Military Academy (USMA), his M.S. at the Naval Postgraduate School, and his Ph.D. at Clemson University. He has taught mathematics at USMA and now at Francis Marion University, where he is professor and chair of mathematics. He has authored two textbooks on mathematical modeling and more than 100 technical and educational articles and presentations. He serves as associate contest director for COMAP's Mathematical Contest in Modeling (MCM) and as contest director for the new High School Mathematical Contest in Modeling (HiMCM). His interests include applied mathematics, optimization (linear and nonlinear), mathematical modeling, statistical models for medical research, and computer simulations.

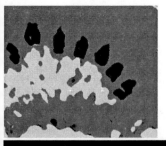

AUTHORS:
Lisa Pike
(Biology)

William P. Fox
(Mathematics)
bfox@fmarion.edu

Francis Marion University
Florence, SC

EDITOR:
Richard D. West
Francis Marion University
Florence, SC

CONTENTS

Survival of Early Americans

MATHEMATICS CLASSIFICATIONS:
College Algebra, Precalculus

DISCIPLINARY CLASSIFICATIONS:
Demography

PREREQUISITE SKILLS:
1. Elementary functions
2. Exponential functions
3. Graphs

PHYSICAL CONCEPTS EXAMINED:
Population dynamics

COMPUTING REQUIREMENT:
Graphing calculator or computer

UMAP/ILAP Modules 2002–2003: Tools for Teaching, 91–104. Reprinted from *The UMAP Journal* 23 (2) (2002) 149–162.

Contents

Setting the Scene

Demography is the study of population dynamics—how populations grow and decline. The worldwide human population is experiencing a population growth phase and presently is increasing at an exponential rate. Today, human population is approximately 6.1 billion people, and it is expected to double in about 35 years (much faster than 600 or 1,000 years ago). We are unsure of the carrying capacity of the earth; some scientists fear that we have already reached it, meaning that we may not have the resources (space, clean water, food etc) for the additional people entering our global population every day.

Populations grow as more members are added, through births and immigration. Populations decline as members are deleted, through deaths and emigration. Stable populations have a balance between growth rates and decay rates (birth and immigration rates and death and emigration rates). So a stable population has a zero growth rate. One way to learn about populations of living organisms is to examine age-specific rates of mortality and reproduction. Because accumulated age-specific information is often presented in life tables, life table analysis is a common component of courses in ecology and population biology. In this project, we perform life-table analysis.

Results can be presented in the form of a *survivorship curve*, which traces the decline in number over time of a group of individuals born at the same time (a *cohort*). It can be thought of as the probability of an individual surviving to various ages—the *life expectancy vs. maximum life span*. For example, the American robin (*Turdus migratorius*) can live to be 7 years old, but the probability of a newly hatched robin doing so is less than 1%. Many robins live only a year or two. Their life expectancy is 1 to 2 years, but their maximum life span is 7 years.

To determine age-specific mortality and survivorship curves for a population, ecologists follow a *cohort*, a group of individuals born within the same time interval. For example, ecologists might follow a cohort for a year, 5 years, or a month, depending on the species that they are looking at. The cohort is followed until all members of the cohort are dead. The gender and age at death are recorded for each individual. The ecologists find that each species has a characteristic life span, with few reaching the maximum age.

The survival rate or *life expectancy* of human populations has increased significantly in the past 100 to 300 years, due to improved nutrition, preventive medicine, lifestyle changes, improved sewage control and hygiene, and new technologies. In the early days of the Roman Empire, life expectancy at birth was only 22 years. In America in 1900, the life expectancy was 48 years; in 1998, it was 76 years (mostly due to a decrease in infant mortality). In 1998, a man in Italy lived to a ripe old age of 126 (the maximum human life span). Survival rates are up and mortality rates, especially infant mortality rates, are down; these factors increase the population growth.

There are three typical population survivorship curves: Type I, Type II, and Type III (**Figure 1**). Humans have a Type I growth rate, with low infant mortality and a high probability of living until you are old (at which time the probability of death increases).

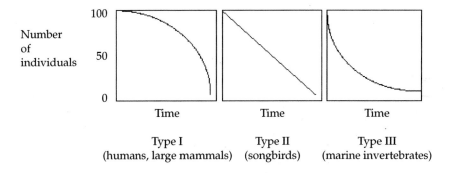

Figure 1. Typical population survival curves.

Situation 1: Survival of Early Americans

In this project, we make survivorship curves for humans living in the 1800s and humans living in the 1900s. We construct a life table from these data using the attached data sheet. Then we determine values for the number of individuals who would have been alive in each interval of ten years. Also, we determine the number of individuals who died during each interval. This curve is the opposite of a survivorship curve and is called a *mortality curve*. A survivorship curve is prepared by plotting the logarithm of the number of survivors against age.

Requirements

Part 1.

1. Complete the tables for males and females 1810–1820 and 1910–1920.

2. Compute from each table the

   (a) AVG Life Expectancy for men

   (b) AVG Life Expectancy for women

   (c) MAX Life Span for men

   (d) MAX Life Span for women

Part 2.

3. Graph log(% survivorship) over time.

4. Compare your graph to the survivorship curves. What kind of survivorship curve did you find for your data (Type I, II, or III)? Why?

5. What does this reflect about males and females in the early 1800s and the early 1900s?

Part 3.

6. What was the average life expectancy for people living in the 1800s? for people living in the 1900s?

7. What was the maximum life span for people living in the 1800s? for people living in the 1900s?

Part 4.

8. Did your data show a difference in age at death between males and females? For which cohort? Why do you think this happened?

Part 5.

9. If a third of the world population is now below the age 15, what effect will this age distribution have on the growth rate of the human population?

10. What sort of humane recommendations would you make to encourage this age group to limit the number of children they plan to have?

Data for Requirements 1–5

Table 1.

Data Table (born 1810–1820): MALES.

Age interval	Males died	Cumulative	Survive	% Survive	% Mortality	$\log_{10}$ of % Survive
0						
0–9	9					
10–19	3					
20–29	7					
30–39	12					
40–49	17					
50–59	26					
60–69	65					
70–79	112					
80–89	67					
90–99	11					
100–109 110 +	4 0					
Total People	333					

Table 2.

Data Table (born 1810–1820): FEMALES.

Age interval	Females died	Cumulative	Survive	% Survive	% Mortality	$\log_{10}$ of % Survive
0						
0-9	4					
10-19	4					
20-29	8					
30-39	10					
40-49	13					
50-59	28					
60-69	41					
70-79	90					
80-89	65					
90-99	11					
100-109	1					
110 +	0					
Total People	275					

Table 3.

Data Table (born 1910–1920): MALES.

Age interval	Males died	Cumulative	Survive	% Survive	% Mortality	$\log_{10}$ of % Survive
0–9	0					
10–19	6					
20–29	2					
30–39	5					
40–49	13					
50–59	19					
60–69	40					
70–79	66					
80–89	144					
90–99	28					
100–109	3					
110+	0					
Total People	326					

Table 4.

Data Table (born 1910–1920): FEMALES.

Age interval	Females died	Cumulative	Survive	% Survive	% Mortality	$\log_{10}$ of % Survive
0–9	1					
10–19	3					
20–29	1					
30–39	4					
40–49	11					
50–59	17					
60–69	36					
70–79	78					
80–89	98					
90–99	24					
100–109	5					
110+	1					
Total People	279					

Situation 2: Age Structure and Survivorship

Populations, whether animal or plant, vary in their proportions of young and old individuals. Time units are used to describe ages but could be in hours, days, weeks, months, or years. Sometimes they can be grouped into classes: nestling, juvenile, subadult, and adult. The proportions collected into the categories are referred to as *age structure* or *age distribution* data. We desire some statistical measures for these ages. In this example, there are 13 categories for age.

Requirements

Part 1.

1. Using our data from the tables produced in Situation 1, compute the following descriptive statistics for ages of males and females in the early 1800s and 1900s. (Use midpoints of the 13 categories for the value of the counts in each class.)

   (a) mean

   (b) median

   (c) mode

   (d) standard deviation

   (e) variance

   (f) range

   (g) coefficient of skewness $SK = 3 \left(\dfrac{\text{mean} - \text{median}}{\text{standard deviation}} \right)$.

2. Interpret the meaning of each of these descriptive statistics.

Part 2.

3. Construct a histogram for each data set. What information is being displayed?

4. Construct boxplots for males and females. Explain the interpretation of the results of the boxplots.

Title: Survival of Early Americans Situation

Sample Solution

Situation 1

Part 1

1. See the tables on the following pages.

2. a) AVG Life Expectancy for men : $22665/333 = 68.06$ years in the 1800s and $24710/326 = 75.79$ years in the 1900s.

  b) AVG Life Expectancy for women : $18845/275 = 68.52$ years in the 1800s and $21085/279 = 75.57$ years in the 1900s.

  c) MAX Life Span for men 100–109 in both.

  d) MAX Life Span for women 100–109 in the 1800s and 110+ in the 1900s.

Part 2

3. See **Figures S1–S4**.

4. It is type I. This makes sense because the data is human.

5. By looking at the data and the percents in each intervals, it is clear that males and females both increased their life span from the early 1800s to the early 1900s.

Part 3

6. 1800s: 68.3 years; 1900s: 75.7 years. People live about 7.4 years longer in the 1900s.

7. MAX Life Span for men: 100–109 in both epochs.
MAX Life Span for women: 100–109 in the 1800s and 110+ in the 1900s.

Part 4

8. Women live longer than men. The 1900s women live longer than both men and the 1800s women. There are many factors including technology, medicine, gender, and adaptability.

Part 5

9. Since the average life expectancy is now over 75 years, then over time this one-third of the world population will continue to survive for many years (approximately 60 more years). The population growth could cause problems in terms of space, food, water, jobs, etc.

10. Population growth will exceed capacity to produce food and other essentials. If this occurs, then the growth rate will slow and the death rate will climb; we will experience a shorter life expectancy.

Table 1.

Data Table (born 1810–1820): MALES.

Age interval	Males died	Cumulative	Survive	% Survive	% Mortality	$\log_{10}$ of % Survive
0	0	0	333	100	0	2
0–9	9	9	324	97.2973	2.702703	1.988101
10–19	3	12	321	96.3964	3.603604	1.984061
20–29	7	19	314	94.29429	5.705706	1.974485
30–39	12	31	302	90.69069	9.309309	1.957563
40–49	17	48	285	85.58559	14.41441	1.932401
50–59	26	74	259	77.77778	22.22222	1.890856
60–69	65	139	194	58.25826	41.74174	1.765357
70–79	112	251	82	24.62462	75.37538	1.39137
80–89	67	318	15	4.504505	95.4955	0.653647
90–99	11	329	4	1.201201	98.7988	0.079616
100–109	4	333	0	0	100	0
110+	0	333	0	0	100	0
TOTAL	333					

Table 2.

Data Table (born 1810–1820): FEMALES.

Age interval	Females died	Cumulative	Survive	% Survive	% Mortality	$\log_{10}$ of % Survive
0	0	0	275	100	0	2
0-9	4	4	271	98.54545	1.454545	1.993637
10–19	4	8	267	97.09091	2.909091	1.987179
20–29	8	16	259	94.18182	5.818182	1.973967
30–39	10	26	249	90.54545	9.454545	1.956867
40–49	13	39	236	85.81818	14.18182	1.933579
50–59	28	67	208	75.63636	24.36364	1.878731
60–69	41	108	167	60.72727	39.27273	1.783384
70–79	90	198	77	28	72	1.447158
80–89	65	263	12	4.363636	95.63636	0.639849
90–99	11	274	1	0.363636	99.63636	-0.43933
100–109	1	275	0	0	100	0
110+	0	275	0	0	100	0
TOTAL	275					

Table 3.

Data Table (born 1910–1920): MALES.

Age interval	Males died	Cumulative	Survive	% Survive	% Mortality	$\log_{10}$ of % Survive
0	0	0	326	100	0	2
0–9	0	0	326	100	0	2
10–19	6	6	320	98.15951	1.840491	1.991932
20–29	2	8	318	97.54601	2.453988	1.98921
30–39	5	13	313	96.01227	3.98773	1.982327
40–49	13	26	300	92.02454	7.97546	1.963904
50–59	19	45	281	86.19632	13.80368	1.935489
60–69	40	85	241	73.92638	26.07362	1.868799
70–79	66	151	175	53.68098	46.31902	1.72982
80–89	144	295	31	9.509202	90.4908	0.978144
90–99	28	323	3	0.920245	99.07975	-0.0361
100–109	3	326	0	0	100	0
110+	0	326	0	0	100	0
TOTAL	326					

Table 4.

Data Table (born 1910–1920): FEMALES.

Age interval	Females died	Cumulative	Survive	% Survive	% Mortality	$\log_{10}$ of % Survive
0	0	0	279	100	0	2
0–9	1	1	278	99.64158	0.358423	1.998441
10–19	3	4	275	98.56631	1.433692	1.993728
20–29	1	5	274	98.20789	1.792115	1.992146
30–39	4	9	270	96.77419	3.225806	1.98576
40–49	11	20	259	92.83154	7.168459	1.967696
50–59	17	37	242	86.73835	13.26165	1.938211
60–69	36	73	206	73.83513	26.16487	1.868263
70–79	78	151	128	45.87814	54.12186	1.661606
80–89	98	249	30	10.75269	89.24731	1.031517
90–99	24	273	6	2.150538	97.84946	0.332547
100–109	5	278	1	0.358423	99.64158	0
110+	1	279	0	0	100	0
TOTAL	279					

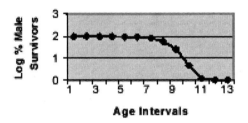

Figure S1. Log of percentage survivorship for males born 1810–1820.

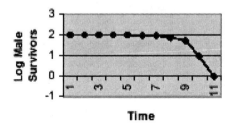

Figure S2. Log of percentage survivorship for females born 1810–1820.

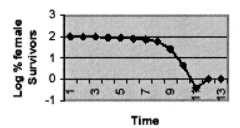

Figure S3. Log of percentage survivorship for males born 1910–1920.

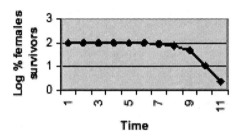

Figure S4. Log of percentage survivorship for females born 1910–1920.

Situation 2

Part 1

1. See **Table S1**.

Table S1.
Statistics for the samples.

Statistic	Male, 1800s	Female, 1800s	Male, 1900s	Female, 1900s
Mean	68.06	68.53	75.798	79.09
Median	75	75	85	75
Mode	75	75	85	85
Std Deviation	19.21	18.84	16.26	18.5
Variance	369.02	354.95	264.39	342.25
Range	0–105	0–105	0–115	0–105
SK	−1.08	−1.03	−1.69	−0.958

2. Each sample has negative skewness, so the distribution is skewed to the left. The locations of mean, median, and mode support this result.

Part 2

3. See **Figures S5–S8**. Histograms provide information about the shape (symmetric or skewed) of the distribution.

4. See **Figure S9**. Boxplots plot the 5-number summary (median, quartiles, and range) to view symmetry, skewness, and outliers. All plots are negatively skewed but are close to the same shape. The five number summaries for these four data sets are extremely similar.

About the First Author ...

Lisa Pike is an assistant professor of biology and has been teaching at Francis Marion University (FMU) for 10 years. She teaches Freshmen Biology, Environmental Science, and Marine Biology. Ms. Pike is also the FMU liaison for the Sustainable Universities Initiative and the faculty advisor for Ecology Club. She was born in Rochester NY; she received a B.S. In Biology from Binghamton University in 1989 and an M.S. in Marine Biology in 1991 from the University of North Carolina at Wilmington. After completing her master's degree, Ms. Pike was selected as a Rotary Scholar and spent a year at James Cook University of North Queensland, Australia.

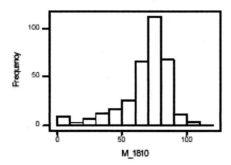

Figure S5. Histogram of ages at death for males in 1800s cohort.

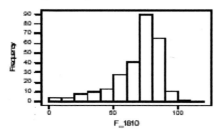

Figure S6. Histogram of ages at death for females in 1800s cohort.

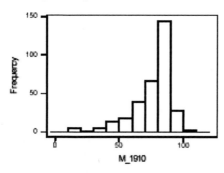

Figure S7. Histogram of ages at death for males in 1900s cohort.

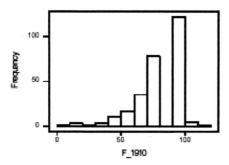

Figure S8. Histogram of ages at death for females in 1900s cohort.

Figure S9. Minitab output for boxplots.

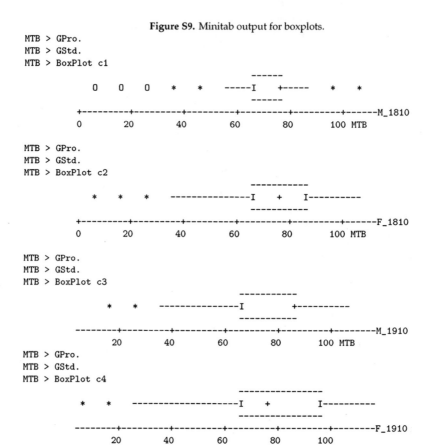

```
MTB > GPro.
MTB > GStd.
MTB > BoxPlot c1
```

```
MTB > GPro.
MTB > GStd.
MTB > BoxPlot c2
```

```
MTB > GPro.
MTB > GStd.
MTB > BoxPlot c3
```

```
MTB > GPro.
MTB > GStd.
MTB > BoxPlot c4
```

... and About the Second Author

Dr. William P. Fox received his B.S. degree from the United States Military Academy (USMA), his M.S. at the Naval Postgraduate School, and his Ph.D. at Clemson University. He has taught mathematics at USMA and now at Francis Marion University, where he is professor and chair of mathematics. He has authored two textbooks on mathematical modeling and more than 100 technical and educational articles and presentations. He serves as associate contest director for COMAP's Mathematical Contest in Modeling (MCM) and as contest director for the new High School Mathematical Contest in Modeling (HiMCM). His interests include applied mathematics, optimization (linear and nonlinear), mathematical modeling, statistical models for medical research, and computer simulations.

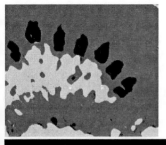

AUTHORS:

John R. Liukkonen
(Mathematics)
jrl@math.tulane.edu

Laura J. Steinberg
(Civil and Environmental
  Engineering)

Tulane University
New Orleans, LA

EDITOR:

David C. Arney
College of St. Rose
Albany, NY

CONTENTS

Chloroform Alert!

University Students Exposed to Unhealthy Air

MATHEMATICS CLASSIFICATIONS:
Mathematical Probability, Multivariable Calculus

DISCIPLINARY CLASSIFICATIONS:
Environmental Engineering

PREREQUISITE SKILLS:
1. Computing means and variances of random variables
2. Understanding partial derivatives
3. Finding Taylor series approximations
3. Understanding the Central Limit Theorem

PHYSICAL CONCEPTS EXAMINED:
1. Advection
2. Diffusion
3. Steady-state modeling
4. Simulation of contaminant transport

COMPUTING REQUIREMENT:
Graphing calculator or computer

UMAP/ILAP Modules 2002–2003: Tools for Teaching, 105–124. Reprinted from *The UMAP Journal* 23 (2) (2002) 163–182.

Contents

Setting the Scene

AAA Chemical Inc. operates two chemical plants in the town of Bayou St. Edwards, home of Dixieland State University. New pollution monitoring equipment has shown elevated concentrations of chloroform in the air over the university. These concentrations are considered unhealthy, and the student Engineering Society has formed a study group to review the problem. They have come up with a number of possible ways to reduce the pollution, including relocating the plants, making the smokestacks taller, and adding air pollution control equipment to the plants. They want to analyze the costs and compute the reduction in chloroform concentration that is expected from each possible solution. You are asked to help them with the project by completing Parts 1, 2, 3, and 4 below.

- In Part 1, you learn how to model chloroform concentrations in the atmosphere in the presence of diffusion and turbulence.

- In Part 2, you add the effects of wind to the model.

- Part 3 shows how the model can be simplified in the case of steady-state conditions (no change in chloroform emissions over time), and how the chloroform emissions from the two chemical production plants can be added together to calculate the combined effect of the two plants on air quality.

- In Part 4, you apply the simplified model to Dixieland University. You are asked to calculate concentrations of chloroform in the air over the university and to choose between various engineering remedies.

Part 1: Diffusion and Turbulence via Discrete Approximation

Prevailing winds carry pollutants from several nearby smokestacks to the university campus. The students want to calculate the levels of chloroform over the campus produced by the smokestacks under various conditions.

The students began with a review of the literature on the dispersion of air pollutants from factory stacks. They learned that to find the chloroform levels, they must use the principle of superposition. That is, they must first calculate the effects from each stack individually, and then add together the effects of all the smokestacks. In fact, they learned that even to find the effects of continuous emission from a single smokestack, superposition would be necessary. For this, they would take an integral in time of so-called *fundamental solutions* to the problem. So their overall strategy is to deal with dispersion in as simple a situation as possible and then use superposition to build solutions for the more complex situation that they face.

The students also learned that dispersion is governed by the effects of advection, diffusion, and turbulence. The effects of diffusion and turbulence can be represented as a *random walk*. A random walk is a mathematical model of a particle moving randomly, one step at a time, around $\mathbf{R}$ or $\mathbf{R}^n$. We use the random walk model to do two things:

- First, we derive the partial differential equation (PDE) for the concentration of molecules undergoing diffusion and turbulence.

- We then find the fundamental solutions to the PDE. These are the concentrations resulting from an initial unit mass at a single point.

In later parts, we modify our solutions to take account of the effects of advection, and see how to use superposition to find the concentrations corresponding to more complex situations.

We begin with a mass of particles located at discrete positions, able to move at discrete times, and subject to diffusion and turbulence. To simplify matters, we assume for now that the particle motion is taking place in one dimension. So consider a mass M_0 of particles distributed on a fine grid on the real line $\mathbf{R}$. Though the total mass M_0 is finite, we assume that the actual number of particles is virtually infinite and that all particles have the same mass. Let

$$\ldots, -n\Delta x, -(n-1)\Delta x, \ldots, -\Delta x, 0, \Delta x, \ldots, n\Delta x, \ldots$$

denote the possible particle positions, where Δx is very small. At the same time, suppose that the particles can move only at times

$$t = 0, \Delta t, 2\Delta t, \ldots, n\Delta t, \ldots,$$

where Δt is also very small. For each $x = m\Delta x$ and $t = n\Delta t$, let $M(x, t)$ be the mass of particles located at position x at time t. Thus, for each t the sum over x of $M(x, t)$ is M_0.

Let $C(x, t)$ be the concentration of mass (mass per unit length) at point x at time t. In our discrete approximation, we have $C(x, t) = M(x, t)/\Delta x$. When we obtain $C(x, t)$ for all x and t by passing to a limit as $\Delta x \to 0$, then for each t the mass in the interval $[a, b]$ at time t is given by

$$\int_a^b C(x, t)dx. \tag{1}$$

We model the particle motion as a simple symmetric random walk. That is, at each time t each particle moves a distance Δx randomly either to the right or to the left. The direction of motion is independent of all past motion and equally likely to be right or left. (Thus it is impossible for the particle to remain at rest in our model. This assumption is not restrictive and keeps things simple.) Given these assumptions, we have the equations

$$M(x, t + \Delta t) = \tfrac{1}{2}M(x - \Delta x, t) + \tfrac{1}{2}M(x + \Delta x, t)$$
$$C(x, t + \Delta t) = \tfrac{1}{2}C(x - \Delta x, t) + \tfrac{1}{2}C(x + \Delta x, t) \tag{2}$$

for every x, t in our grids.

Requirement 1. Use **(2)** to express $C(x, t + \Delta t) - C(x, t)$ as a linear combination of $C(x + \Delta x, t), C(x, t)$, and $C(x - \Delta x, t)$.

We now use your result from **Requirement 1** to obtain a relation between $\partial C(x, t)/\partial t$ and $\partial^2 C(x, t)/\partial x^2$ for the case of particles continuously distributed along **R** moving at all positive times. To find such a relation, it is possible to seek out approximating difference quotients for the derivatives we want, but we choose to use Taylor approximations and see what falls out. Of course, the difference-quotient approach properly executed gives the same relationship.

Requirement 2.

   (a) For small Δt, approximate $C(x, t + \Delta t) - C(x, t)$ by a first-order polynomial in Δt. Use a first order Taylor expansion for C in the second variable with base point (x, t). Neglect the remainder term. Your answer should involve $\partial C(x, t)/\partial t$.

   (b) For small Δx, approximate the linear combination of $C(x + \Delta x, t), (Cx, t)$, and $C(x - \Delta x, t)$ you got in **Requirement 1** by a second-order polynomial in Δx. Use a second order Taylor expansion for C in the first variable, again with base point (x, t). Neglect the remainder term. Your answer should involve $\partial^2 C(x, t)/\partial x^2$.

   (c) Set your answer in 2(a) equal to your answer in 2(b). Now assume that $\Delta t = c(\Delta x)^2$. (We will see later that the standard deviation of the position in a random walk is proportional to the square root of the time elapsed—i.e., a typical x displacement is proportional to $\sqrt{t}$.) So divide the left-hand side of your equation by Δt, and the right-hand side by $c(\Delta x)^2$. Obtain an equation relating $\partial C(x, t)/\partial t$ and $\partial^2 C(x, t)/\partial x^2$.

If we set $k_1 = (2c)^{-1}$ in your equation, we now have the following partial differential equation for the concentration $C(x, t)$:

$$\frac{\partial C(x, t)}{\partial t} = k_1 \frac{\partial^2 C(x, t)}{\partial x^2}. \tag{3}$$

The constant k_1 is called the *mass diffusivity*; its value depends on the particular situation.

To determine the function $C(x, t)$ uniquely for all (x, t), we not only need to know that it satisfies (3) but we also need additional information on C. The usual requirement is specification of an initial distribution $C(x, 0)$ for all x, but our situation is unusual. Rather than an initial distribution function for C over an interval of x, we must deal with an initial concentration of mass at a single point x_0 (the location of a smokestack). Can we specify an initial concentration function for this situation?

Since we have positive unit mass at a single point and no initial mass elsewhere, our concentration function—the mass "density"—must be infinite at x_0 and zero elsewhere, and its integral over all x must yield the unit mass. Thus we must work with a "function" that is zero everywhere except a single point, where it is infinite, and its integral is 1.

Such a "function" is called a *Dirac delta function*. It is not really a function, but everyone agrees that it is a very useful idea, so everyone agrees to treat it as a function. The standard notation for a Dirac delta function concentrated at x_0 is $\delta_{x_0}(x)$. Its basic properties are:

- $\delta_{x_0}(x) = 0$ for $x \neq x_0$,

- $\delta_{x_0}(x_0) = \infty$, and

- $\int_{-\infty}^{\infty} \delta_{x_0}(x) dx = 1$.

For any function f, we have $\int_{-\infty}^{\infty} f(x)\delta_{x_0}(x) dx = f(x_0)$.

The concentration $C_{x_0}(x, t)$ corresponding to an initial Dirac delta function is called a *fundamental solution* to our problem, because the concentration $C(x, t)$ corresponding to an arbitrary initial distribution $C(x, 0) = f(x)$ can be built up from these fundamental solutions. In fact if $C_{x_0}(x, t)$ solves (3) for each x, t and we have $C_{x_0}(x, 0) = \delta_{x_0}(x)$ for each x_0, x, then $C(x, t) = \int_{-\infty}^{\infty} f(u)C_u(x, t) du$ satisfies (3) and $C(x, 0) = f(x)$ for all x. Here we assume that we can differentiate under integrals.

We now derive the formula for the concentration corresponding to an initial point mass at $x = x_0$; that is, for the fundamental solution corresponding to initial Dirac delta function δ_{x_0}. We denote the concentration that we seek by $C_{x_0}(x, t)$. To work out the formula for $C_{x_0}(x, t)$, we return to the discrete grids that we used above and track the motion of a single particle beginning at x_0 at time $t = 0$. Its position at time $t = 0$ is fixed at x_0, but its position at any positive time $t = n\Delta t$ is random. To derive the probability distribution of its position at time t, we introduce random variables $X_1, X_2, \ldots$. For each k, the random variable $X_k = +1$ if the particle moves to the right at time $t = k\Delta t$ and $X_k = -1$ if the particle moves left at that time. The X_ks are independent and each X_k has distribution $P(X_k = 1) = 0.5$ and $P(X_k = -1) = 0.5$ for each k. The random position $S(t)$ of our particle at time $t = n\Delta t$ can be written

$$S(t) = x_0 + \Sigma_{k=1}^{n} X_k \Delta x.$$

According to the Central Limit Theorem from probability theory, if n is large, the distribution of $S(t)$ is approximately normal. The mean of $S(t)$ is x_0 and the standard deviation of $S(t)$ is $\Delta x \sqrt{n}$.

Requirement 3. Recall from probability theory that for random variables $X_1, \ldots, X_n$ the mean of the sum $X_1 + \cdots + X_n$ is the sum of the means of the X_is. If in addition the $X_1, \ldots, X_n$ are independent, the variance of the sum is the sum of the variances. Use these facts to verify our claims about the mean and standard deviation of $S(t)$.

Requirement 4. Now assume that $\Delta t = (\Delta x)^2/(2k_1) = t/n$. What are the mean and variance of $S(t)$ in terms of t? Your answer should depend on t but not on Δx or n.

Since the distribution of $S(t)$ becomes normal as $n \to \infty$ and we know the mean and variance of $S(t)$, we can write down a limiting density for $S(t)$ as a function of x.

Requirement 5. Write down the formula for the limiting density of $S(t)$; that is, the normal density with mean x_0 and the standard deviation you found in **Requirement 4**.

If we now return to our assumptions that the number of mass particles is virtually infinite and the particles all have equal mass, and assume also that the total mass is 1, then the limiting density that you wrote down in **Requirement 5** is the mass density in x at time t. In other words, it is the desired solution $C_{x_0}(x, t)$.

Requirement 6. Verify that $C_{x_0}(x, t)$ satisfies equation **(3)**.

The remaining point to check is that $C_{x_0}(x, 0)$ is a Dirac delta function at x_0. We point out $C_{x_0}(x, t)$ is a normal density with mean x_0 and standard deviation $\sqrt{2k_1 t}$. Hence $C_{x_0}(x, t)$ always integrates over all x to 1, and for t very close to zero, $C_{x_0}(x, t)$ is concentrated near x_0 and nearly infinite at x_0. So we believe that $C_{x_0}(x, 0)$ is a Dirac delta function at x_0. Finally, we ask you to take on faith that there can be only one positive solution to **(3)** whose initial distribution is the Dirac delta function at x_0.

Part 2: Advection via Discrete Approximation

In this section, we consider the effects of wind (advection) on our problem. For simplicity, we assume that the wind blows constantly in the same direction. Just as we used discrete approximations to investigate diffusion and turbulence, we do the same for advection. Let us again consider the fine grid as in Part 1, but we neglect the effects of diffusion and turbulence and concentrate on advection for now. We assume that we have a wind whose velocity in the positive direction is c. We model this advective effect by the equations

$$M(x, t + \Delta t) = M(x - c\Delta t, t)$$
$$C(x, t + \Delta t) = C(x - c\Delta t, t). \tag{4}$$

The right-hand side of the first equation represents the mass with position $x - c\Delta t$ at the current time t. At time $t + \Delta t$, is be the mass at position x, which mass is represented by the left hand side of the first equation. Hence the first equation; the second equation is simply the first equation divided by Δx.

Requirement 1. Use **(4)** to express $C(x, t + \Delta t) - C(x, t)$ as a linear combination of $C(x - c\Delta t, t)$ and $C(x, t)$.

Requirement 2. Use your result in **Requirement 1** to obtain a relation between $\partial C(x, t)/\partial t$ and $\partial C(x, t)/\partial x$. As in **Requirement 2** of **Part 1**, you should approximate both sides by Taylor polynomials for small Δt, and then divide through by Δt. This time, you should use first-order polynomials in Δt on both sides.

You should have obtained for $C(x, t)$ the equation

$$\frac{\partial C(x, t)}{\partial t} + c\frac{\partial C(x, t)}{\partial x} = 0. \tag{5}$$

We need to put the effects of diffusion and turbulence and advection together. For the governing equation, the standard assumption—the wisdom of the field, if you will—is that the two effects are additive; thus, for the concentration $D(x, t)$ undergoing advection as well as diffusion and turbulence, we arrive at

$$\frac{\partial D(x, t)}{\partial t} + c\frac{\partial D(x, t)}{\partial x} = k_1 \frac{\partial^2 D(x, t)}{\partial x^2}. \tag{6}$$

A good guess for the fundamental solution of this equation is simply to take the fundamental solution C_{x_0} that we found in **Part 1** and let it drift with velocity c:

$$D_{x_0}(x, t) = C_{x_0}(x - ct, t). \tag{7}$$

Requirement 3. Verify that $D_{x_0}(x, t)$ solves **(6)**. (Hint: Express the derivatives of D_{x_0} in terms of the derivatives of C_{x_0}, and use the fact that C_{x_0} satisfies **(3)**.)

We now summarize your results from **Parts 1** and **2**: A concentration D of mass undergoing diffusion, turbulence and advection satisfies **(6)**. The solution D_{x_0} given by **(7)** is the concentration corresponding to an initial unit mass concentrated at x_0.

We now extend, without proof, the equations and solutions you obtained for one space dimension to three space dimensions. From here on, we denote all concentrations by C.

Let $C_{p_0}(p, t)$ be the concentration of mass at $p = (x, y, z)$ at time t corresponding to a unit mass concentrated at $p_0 = (x_0, y_0, z_0)$ at time $t = 0$. (In three dimensions, concentration is mass per unit volume.) Assume a wind of constant velocity c in the positive x direction. Then we have the equation

$$\frac{\partial C}{\partial t} + c\frac{\partial C}{\partial x} = k_1 \frac{\partial^2 C}{\partial x^2} + k_2 \frac{\partial^2 C}{\partial y^2} + k_3 \frac{\partial^2 C}{\partial z^2}. \tag{8}$$

The fundamental solution C_{p_0} is given by

$$C_{p_0}(p, t) = (4\pi t)^{-3/2}(k_1 k_2 k_3)^{-1/2} \times$$
$$\exp\left(-\left(\frac{(x - x_0 - ct)^2}{4k_1 t} + \frac{(y - y_0)^2}{4k_2 t} + \frac{(z - z_0)^2}{4k_3 t}\right)\right) \tag{9}$$

For the sequel, keep in mind that p and p_0 are position vectors in $\mathbf{R}^3$ and t is the time variable. In particular, in the context of differential equations including variables x, y, z, keep in mind that p is short for (x, y, z) and p_0 is short for (x_0, y_0, z_0).

Part 3: Superposition and Steady State

Having found the fundamental solutions to our diffusion-advection equation, we now want to put these together to deal with mass initially distributed over many locations, and also with mass continuously generated over time, such as would arise from a smokestack. One element that will be missing from our analysis is the effect of the ground—that is, does the ground absorb or reflect the gas in question? However, our smokestacks are elevated, and in this case one can neglect the issues of absorption and reflection with little error, since the bulk of the movement takes place above ground.

To put together more general solutions from fundamental solutions, we make the key observation is that equation **(8)** is linear; thus if C_1 and C_2 both satisfy **(8)**, then so does any linear combination of C_1 and C_2. In particular, if we begin at time $t = 0$ with masses $m_1, m_2, \ldots, m_k$ concentrated at the points $p_1, p_2, \ldots, p_k$ in $\mathbf{R}^3$, then the corresponding concentration C is given by

$$C(p, t) = \Sigma_{i=1}^{k} m_i C_{p_i}(p, t). \tag{10}$$

Recall that the vector p as short for (x, y, z). If on the other hand we begin at time $t = 0$ with mass distributed over $\mathbf{R}^3$ according to the initial concentration $f(x, y, z)$, then the corresponding concentration C is

$$C(p, t) = \int f(u, v, w) C_{(u,v,w)}(p, t) \, du \, dv \, dw. \tag{11}$$

This process of taking linear combinations of fundamental solutions to a linear equation to obtain solutions corresponding to more complex initial conditions is called *superposition*.

Requirement 1. Give the solution to **(8)** corresponding to an initial distribution of 1 unit of mass at $(0, 0, 1)$ and 2 units of mass at $(0, 2, 1)$.

In our application, we need to consider not only mass generated at several points at the same time but also mass generated at the same point at different times. We handle this by superposition in the time variable. Let $C_{p_0}^{t_0}(p, t)$ denote the concentration at point p at time t corresponding to what was a unit mass at p_0 at time t_0. Then

$$C_{p_0}^{t_0}(p, t) = C_{p_0}(p, t - t_0). \tag{12}$$

Requirement 2. Verify that $C_{p_0}^{t_0}$ satisfies **(8)** for each p_0, t_0. (Hint: Relate the derivatives of $C_{p_0}^{t_0}$ to those of C_{p_0} and use the fact that C_{p_0} satisfies **(8)**.) Let C denote the concentration corresponding to a unit of mass generated at p_0 at time $t = 0$ and another unit of mass generated at p_0 at time $t = 1$. Find C.

Requirement 3.

(a) Suppose that starting at time $t = A$, mass is continuously generated at $p = (0, 0, 0)$ according to a mass rate $dM/dt = Q(t)$. Find the concentration $C(x, t)$ by superposition in t. Your answer may be left in the form of an integral. (Hint: For each time s in the interval $[A, t]$, an initial mass $Q(s) ds$ is generated at $(0, 0, 0)$. Each unit of mass generated at $(0, 0, 0)$ at time s contributes $C_{(0,0,0)}^s(p, t)$ to the total concentration at (p, t).)

(b) (optional) Assuming $Q(t) = Q_0$ is constant, show that your integral satisfies

$$\frac{\partial C}{\partial t} + c \frac{\partial C}{\partial x} = k_1 \frac{\partial^2 C}{\partial x_1^2} + k_2 \frac{\partial^2 C}{\partial x_2^2} + k_3 \frac{\partial^2 C}{\partial x_3^2} + Q_0 \delta_{(0,0,0)}(x). \tag{13}$$

Here $Q_0 \delta_{(0,0,0)}(x)$ is a *forcing term* corresponding to continuous generation of mass.

It is possible, given sufficient data on the pollution source, to calculate the concentration $C(x, y, z, t)$ by superposition, as outlined above. However, several simplifying assumptions lead to easier calculations. The first simplifying assumptions are that the rate of emissions is a constant Q_0 and that we have a steady state—that is, the emissions have been going on for some time and the resulting concentrations do not vary with time.

Let us give some justification for this. If in your answer to **Requirement 3a** we let $A \to -\infty$, we get

$$C(p, t) = Q_0 \int_{-\infty}^{t} C_{(0,0,0)}(p, t - s)ds = Q_0 \int_{0}^{\infty} C_{(0,0,0)}(p, r)\, dr \qquad (14)$$

after the change of variable $r = t - s$. Notice that the last integral does not depend on t. Now, to save space and effort, let

$$L = k_1 \frac{\partial^2}{\partial x^2} + k_2 \frac{\partial^2}{\partial y^2} + k_3 \frac{\partial^2}{\partial z^2} - c \frac{\partial}{\partial x}. \qquad (15)$$

Then accepting that we can differentiate under integrals we get, for $C(p, t)$ defined as in **(14)**,

$$LC(p, t) = Q_0 \int_{0}^{\infty} LC_{(0,0,0)}(p, r)\, dr = Q_0 \int_{0}^{\infty} \frac{\partial C_{(0,0,0)}(p, r)}{\partial r}\, dr \qquad (16)$$

since $C_{(0,0,0)}$ satisfies equation **(8)**. But this last integral is equal to $-Q_0 \delta_{(0,0,0)}$ since $C_{(0,0,0)}(p, t) \to 0$ as $t \to \infty$ and $C_{(0,0,0)}(p, 0) = \delta_{(0,0,0)}(p)$. Thus, $C(p, t)$ as defined in **(14)** satisfies

$$c \frac{\partial C}{\partial x} = k_1 \frac{\partial^2 C}{\partial x^2} + k_2 \frac{\partial^2 C}{\partial y^2} + k_3 \frac{\partial^2 C}{\partial z^2} + Q_0 \delta_{(0,0,0)}(p), \qquad (17)$$

which is just equation **(13)** with the time derivative removed.

Further simplification is possible and customary. When the wind is appreciable, the advective effect in the x-direction dominates the diffusion effect in that direction. Therefore we usually drop the $\partial^2 C/\partial x^2$ term out of **(17)** and divide through by c to get

$$\frac{\partial C}{\partial x} = k_2 \frac{\partial^2 C}{\partial y^2} + k_3 \frac{\partial^2 C}{\partial z^2} + \frac{Q_0}{c} \delta_{p_0}(p). \qquad (18)$$

Except for the forcing term, this looks like **(8)** but with 2 space dimensions and with t replaced by x. Since the forcing term does not apply for positive x, we drop it from the equation and make it an initial condition $C(0, y, z) = (Q_0/c)\delta_{(y_0, z_0)}(y, z)$. Adjusting the solution **(9)** to meet our current needs, we get a simplified approximate formula for the steady-state solution:

$$C(x, y, z) = \frac{Q_0}{4\pi x \sqrt{k_2 k_3}} \exp\left[-\frac{c}{4x} \left(\frac{(y - y_0)^2}{k_2} + \frac{(z - z_0)^2}{k_3} \right) \right]. \qquad (19)$$

We have done a lot of simplifying and assuming. It is worthwhile to check up on this work numerically, for which it is helpful to rewrite **(14)** carefully. To keep things simple, let us take $p_0 = (0,0,0)$, let $s = ct$, and let $\gamma = k_1/c$. Then we can write **(14)** as

$$\int_0^\infty \frac{1}{\sqrt{4\pi\gamma s}} \exp\left[-\frac{(x-s)^2}{4\gamma s}\right] \frac{Q_0}{4\pi s\sqrt{k_2 k_3}} \exp\left[-\frac{c}{4s}\left(\frac{y^2}{k_2} + \frac{z^2}{k_3}\right)\right] ds. \qquad (20)$$

Thus in **(14)** we are integrating

$$\frac{Q_0}{4\pi s\sqrt{k_2 k_3}} \exp\left[-\frac{c}{4s}\left(\frac{y^2}{k_2} + \frac{z^2}{k_3}\right)\right] \qquad (21)$$

against

$$\frac{1}{\sqrt{4\pi\gamma s}} \exp\left[-\frac{(x-s)^2}{4\gamma s}\right] \qquad (22)$$

for s ranging from 0 to ∞. We now relate **(19)** to **(14)**.

Requirement 4. (optional)

(a) Use a scientific calculator or computer package to convince yourself that the integral of **(20)** in s from 0 to ∞ is 1, no matter what x or γ are.

(b) Convince yourself with a graphics calculator or computer graphics package that as $\gamma \to 0$, the function in **(22)** becomes a Dirac delta function concentrated at $s = x$.

(c) Substitute $s = x$ in **(21)**. You should get **(19)**, with $p = (0,0,0)$. .

The upshot of **Requirement 4** is

• **(14)** is an average over s of the two dimensional Gaussian densities in the variables y, z given by **(21)**;

• As $k_1/c \to 0$, the mixing function, **(22)** becomes concentrated at $s = x$;

• For $s = x$, **(21)** becomes **(19)**, with $p = (0,0,0)$.

Therefore for small k_1/c **(14)** is well approximated by **(18)**.

The requirements in **Part 4** ask you to carry out various calculations using **(18)**. In those situations, $\gamma = k_1/c$ is not so small, but x is quite large, and this also has an impact: It limits the effective range of the variable s in **(21)** to an interval around x whose length is a relatively small fraction of x, and **(20)** is relatively constant over that range. So we can hope that in this case also that **(14)** is well approximated by **(18)**.

However, rather than ask you to chase through analytic verifications of all these claims, we suggest that you simply compare some answers that you get in **Part 4** by the simplified approximate formula **(18)** to the answers that you would get from the superposition formula **(14)**. This will be part (b) of **Requirement 1** in **Part 4**.

Part 4: Data

Having derived—with your assistance—the equations for calculating the concentration of chloroform in the air due to a stack discharge, the students then collected data on the two plants. They found the following information:

Plant 1:

- $Q_0 = 100$ g/s

- stack height: $z = 25$ m

- Distance from Student Union (parallel to the predominant wind direction): $x = 4000$ m

- Distance from Student Union (perpendicular to the predominant wind direction): $y = 300$ m

Plant 2:

- $Q_0 = 50$ g/s

- stack height: 10 m

- Distance from Student Union (parallel to the predominant wind direction): $x = 1100$ m

- Distance from Student Union (perpendicular to the predominant wind direction): $y = 45$ m

Students in the Environmental Engineering Department have conducted atmospheric diffusion experiments and determined that appropriate values for k_2 and k_3 are 0.45 m^2/s and 0.32 m^2/s, respectively. Also, wind velocities in the area are typically 0.5 m/s.

Requirement 1.

(a) Predict the chloroform concentrations in the ambient air over the Student Union due to the emissions from these plants under the prevailing wind conditions.

(b) Consider the two smokestacks introduced in **Part 4**. For each smokestack use a graphing calculator or computer package to determine intervals where the corresponding integrand in **(14)** is concentrated. Then use numerical integration from a calculator or computer package to evaluate **(14)** for each smokestack. Assume $k_1 = 0.45$. Your answers should be fairly close to those you obtained in part (a).

Requirement 2. Suppose that an atmospheric inversion occurs and a mass of stagnant air settles over the area for several days. In this case, wind velocities drop to approximately 0.05 m/s. Find the resulting chloroform concentrations. Compare these to the concentrations found with the usually prevailing wind conditions. Under which conditions is the air quality healthier?

Requirement 3. The students have proposed several methods for reducing chloroform concentrations in the ambient air over the campus:

1. Increase each stack's height by 8 m.

2. Add pollution control equipment, such as air scrubbers, to the stacks to reduce the mass rate by 35 g/s.

3. Shut down production at Plant 2 and transfer all operations to Plant 1.

4. Move Dixieland State University to Hawaii so that students and faculty can escape the chloroform without interrupting chemical plant operations.

Determine the effect that each of these options would have on ambient air concentrations of chloroform at the Student Union. Discuss the options (including the chemical plants as presently constructed) from the points of view of ambient air quality, cost (estimated), and feasibility. Provide a recommendation as to how the chloroform emissions should be reduced.

Requirement 4. When the students take their recommendations to AAA Chemical Inc., company managers reply that cost considerations should be evaluated more scientifically. The managers supply the students with the following data.

- Cost of increasing stack height: $30,000 fixed cost plus $5,000 per meter (to a maximum stack height of 40 m)

- Cost of pollution control equipment per plant : $Cost = 10,000 + 2,000r^2$, where r is the number of grams/second reduced (to a minimum emission of 2 g/s at each plant) and Cost is in dollars.

- Shut down and transfer operations at either plant: $2,000,000.

- Relocate Dixieland State University: $500 million to rebuild the university in Hawaii, offset by $200 million in sale of Bayou St. Edwards property and $100 million in present worth value of the larger student population eager to attend college in Hawaii.

  (a) Calculate the cost of building each stack to a 40-meter-height. What is the resulting ambient concentration at the Student Union under prevailing wind conditions?

(b) Calculate the cost of reducing emissions at each plant to 2 g/s. What is the resultant ambient chloroform concentration at the Student Union under prevailing wind conditions?

(c) Compute the costs of each of the options in Requirement 3. Is your recommendation from Requirement 3 more expensive than the other options? Would you change your initial recommendation based on these costs? Write a paragraph comparing the cost and effectiveness of each option.

Requirement 5.

(a) The cost of transferring the chemical plant operations to a single plant is $2,000,000. By how much would emissions at each plant be reduced if $1,000,000 is spent on pollution control equipment at each plant? What would the resulting ambient concentrations at the Student Union be?

(b) What is the optimal allocation of the $2,000,000 for pollution control equipment between the two plants? You should base your answer on the allocation that provides the lowest ambient chloroform concentration at the Student Union.

Sample Solutions

Part 1

Requirement 1.

$$C(x, t + \Delta t) - C(x, t) = \tfrac{1}{2}C(x + \Delta x, t) - C(x, t) + \tfrac{1}{2}C(x - \Delta x, t). \quad \textbf{(S1)}$$

Requirement 2. Using a first-degree Taylor expansion in variable t, the left-hand side of **(S1)** can be approximated by $\partial C(x, t)\Delta t$. Using second-degree Taylor expansions in the variable x, the right-hand side of **(S1)** can be approximated by

$$\tfrac{1}{2}\left(C(x, t) + \frac{\partial C(x, t)}{\partial x}\Delta x + \frac{\partial^2 C(x, t)}{\partial x^2}\frac{\Delta x^2}{2}\right)$$
$$- C(x, t) + \frac{1}{2}\left(C(x, t) - \frac{\partial C(x, t)}{\partial x}\Delta x + \frac{\partial^2 C(x, t)}{\partial x^2}\frac{\Delta x^2}{2}\right).$$

Simplifying the right-hand Taylor expansion, assuming $\Delta t = c(\Delta x^2)$, and then dividing through by Δt, we get

$$\frac{\partial C(x, t)}{\partial t} = \frac{1}{2c}\left(\frac{\partial^2 C(x, t)}{\partial x^2}\right). \quad \textbf{(S2)}$$

Requirement 3. $E[S(t)] = x_0 + \Sigma_{k=1}^{n}E(X_k)\Delta x = x_0$ since $E[X_k] = 0$ for each k. Also, the variance $\text{Var}\,[S(t)] = \Sigma_{k=1}^{n}\text{Var}\,(\Delta x X_k) = n(\Delta x)^2$. So the standard deviation of $S(t)$ is $\Delta x\sqrt{n}$.

Requirement 4. $E[S(t)] = x_0$ and $\text{Var}\,(S(t)) = n(\Delta x)^2 = 2k_1 t$.

Requirement 5. $C_{x_0}(x, t) = (4\pi k_1 t)^{-1/2}\exp\left[-(x - x_0)^2/4k_1 t\right].$

Requirement 6.

$$\frac{\partial C_{x_0}(x, t)}{\partial t} = (4\pi k_1)^{-1/2}\left(-\frac{1}{2t^{3/2}} + \frac{(x - x_0)^2}{4k_1 t^{5/2}}\right)\exp\left[-\frac{(x - x_0)^2}{4k_1 t}\right]. \quad \textbf{(S3)}$$

But

$$\frac{\partial^2 C_{x_0}(x, t)}{\partial x^2} = (2\pi k_1 t)^{-1/2}\left(-\frac{1}{2k_1 t} + \frac{(x - x_0)^2}{4(k_1 t)^2}\right)\exp(-\frac{(x - x_0)^2}{4k_1 t}). \quad \textbf{(S4)}$$

It follows immediately that C_{x_0} satisfies **(3)**.

Part 2

Requirement 1. $C(x, t + \Delta t) - C(x, t) = C(x - c\Delta t, t) - C(x, t)$.

Requirement 2. Approximating the left-hand side of the equation in **Requirement 1** by a first-order Taylor polynomial in the variable t and the right-hand side of that equation by a first-order Taylor polynomial in the variable x, we get $(\partial C(x, t)/\partial t)\Delta t = -c(\partial C(x, t)/\partial x)\Delta t$. Dividing through by Δt, we get

$$\frac{\partial C(x, t)}{\partial t} + c\frac{\partial C(x, t)}{\partial x} = 0. \tag{S5}$$

Requirement 3. The key here is to use our results from **Part 1**, and not repeat all the differentiation. From the defining equation **(7)**, we get $\partial D_{x_0}(x, t)/\partial t = -c\partial C_{x_0}(x - ct, t)/\partial x + \partial C_{x_0}(x - ct, t)/\partial t$. Also from **(7)**, we have

$$\partial D_{x_0}(x, t)/\partial x = \partial C_{x_0}(x - ct, t)/\partial x$$

and

$$\partial^2 D_{x_0}(x, t)\partial x^2 = \partial^2 C_{x_0}(x - ct, t)/\partial x^2.$$

Putting all these together and using **(5)**, we get that D_{x_0} satisfies **(6)**.

Part 3

Requirement 1. $C_{(0,0,1)} + 2C_{(0,2,1)}$.

Requirement 2. Each derivative of $C_{p_0}^{t_0}$ involved in this equation is the corresponding derivative of C_{p_0} evaluated at $(p, t - t_0)$. Since C_{p_0} satisfies **(8)**, so does $C_{p_0}^{t_0}$. For the conditions given here, $C = C_{p_0}^0 + C_{p_0}^1$.

Requirement 3.

a) $C(x, t) = \int_A^t Q(s)C_{(0,0,0)}^s(p, t)ds$.

b) For convenience, define L as in equation **(15)**. Then accepting that we can differentiate under integrals, we get

$$\frac{\partial C}{\partial t} = Q_0 C_{(0,0,0)}^t(p, t) + Q_0 \int_A^t \frac{\partial C^s}{\partial t(p, t)}\, ds$$

$$= Q_0 C_{(0,0,0)}(p, 0) + \int_A^t L C_{(0,0,0)}^s(p, t)\, ds$$

$$= Q_0 C_{(0,0,0)}(p, 0) + LC.$$

Requirement 4.

a) It is possible to show that this integral is 1, as Victor Moll has shown us. However, plugging in various values of x and γ will also convince us of this.

b) Use a standard graphing calculator, such as the TI-82, adjusting the window.

c) Obvious.

Part 4

Requirement 1.
 a) Plant 1: 9.52×10^{-6} g/m^3. Plant 2: 5.52×10^{-3} g/m^3. Total ambient air concentration: 5.52×10^{-3} g/m^3.
 b) Plant 1: **(14)** integrand concentrated in r on [7500,8500]; $C = 9.6 \times 10^{-6}$. Plant 2: **(14)** integrand concentrated in r on [1900,2500]; $C = .0055$.

Requirement 2. Plant 1: 2.79×10^{-3} g/m^3. Plant 2: 9.02×10^{-3} g/m^3. Total ambient air concentration: 12.81×10^{-3} g/m^3.

Requirement 3.
 Option 1: Plant 1: 9.10×10^{-6} g/m^3 Plant 2: 5.09×10^{-3} g/m^3. Total ambient air concentration: 5.09×10^{-3} g/m^3.
 Option 2: Plant 1: 6.19×10^{-6} g/m^3 Plant 2: 1.66×10^{-3} g/m^3. Total ambient air concentration: 1.66×10^{-3} g/m^3.
 Option 3: Combined plant at Plant 1 location: 1.42×10^{-5} g/m^3.
 Option 4: 27.07×10^{-3} g/m^3.
 Option 5: 0 g/m^3, assuming that there are no chloroform-emitting plants in Hawaii within transport distance of the University.

Costs and Feasibility:
 Option 1: High-cost option. Must add height to existing structure. Makes little difference in ambient concentrations.
 Option 2: Lower capital costs but will increase operating and maintenance costs. Very feasible—the common practice, in fact. Best for the overall environment since total emissions are reduced.
 Options 3 & 4: May actually be less expensive to operate in the long run since operations are centralized. May be infeasible if the company had a good reason to separate the plants to begin with. Do not reduce the total amount emitted; but if all operations are moved to Plant 1, the concentration at the Student Union is greater reduced.
 Option 5: Costs are very high; and the project, while tempting, is probably infeasible.
 Existing project: the default choice. Obviously feasible, presumably the preferred choice from a cost perspective.

Requirement 4.
 a.) $105,000 for Plant 1, $180,000 for Plant 2. Concentration due to Plant 1: 8.65×10^{-6} g/m^3. Concentration due to Plant 2: 3.23×10^{-3} g/m^3.
 b.) $19.21 million for Plant 1, $4.62 million for Plant 2. Concentration due to Plant 1: 1.90×10^{-7} g/m^3. Concentration due to Plant 2: 2.21×10^{-4} g/m^3.
 c.) Option 1: $70,000 at each plant. Option 2: $246,00 0 at each plant. Option 3: $2,000,000. Option 4: $2,000,000. Option 5: $200 million.

Requirement 5.

a.) 22.25 g/s. Concentration due to Plant 1: $7.38 \times 10^{-6} g/m^3$. Concentration due to Plant 2: $3.03 \times 10^{-3} g/m^3$.

b.) Optimal allocation would be to use the entire \$2,000,000 to reduce emissions at Plant 2.

Title: Chloroform Alert!

Notes for the Instructor

This ILAP introduces the students to principles of fate and transport modeling and provides a lively application of their (prior) knowledge of elementary probability. The subject matter of the project is drawn from environmental engineering and involves calculating the ambient air concentrations of chloroform which result from air emissions from two chemical plants. Engineering concepts introduced include advective and diffusive transport of contaminants.

The major probability concept developed is the elementary theory of simple symmetric random walks. Required probability background is awareness of the Central Limit Theorem for independent identically distributed random variables and the ability to calculate the mean and variance of a linear combination of independent random variables. These concepts and skills can actually be developed by the instructor provided the students have seen random variables, means and variances of sums of independent random variables, and the Central Limit Theorem. In other words, it is previous exposure rather than a high level of proficiency that is required.

The other mathematical concepts required are partial derivatives and first and second-degree Taylor approximations.

Part 1

The students are led through a derivation of the diffusion equation in one dimension from a discrete approximation. They are led to the fundamental solutions to this equation by using random walks and the Central Limit Theorem, and the students finally verify the solutions.

Part 2

An equation for advective motion in one dimension is developed from the same discrete approximation as in **Part 1**. The equations for the combined effects of diffusion and advection are given, and the students verify a proposed solution. The three-dimensional analogues are stated.

Part 3

The principles of superposition in time and space are stated and used to build up more general solutions to the equations from **Part 2**. Students are asked to carry out some easy discrete and continuous superpositions. The students are asked to give the steady-state equation. After some simplifying assumptions, the students develop the fundamental solutions to the simplified steady-state equation. Also, an integral formula is developed for the solution of the unsimplified steady state equation, and the students are invited to check via graphing calculators/packages that the explicit solution to the simplified equation approximates the integral formula well.

Part 4

The engineering application of the results derived and methods introduced in **Parts 1–3** is developed in **Part 4**. We apply analytical solutions to the governing differential equations and numerical differentiation to determine the current ambient air concentration of chloroform and the ambient air concentrations that would result if certain changes in the plants were made. Then the students are asked to evaluate possible solutions to the problem of unhealthy concentrations of chloroform in the air from the standpoint of feasibility and effectiveness. Finally, the students are presented with cost functions for each solution and asked to compare the cost for implementing each one.

About the Authors

John Liukonnen received his B.A. in mathematics from Harvard and his Ph.D. in mathematics from Columbia. He an associate professor of mathematics at Tulane University, where he has been since 1970. He works in statistics and statistics education.

Laura J. Steinberg holds a B.S.E. in Civil and Urban engineering from the University of Pennsylvania and a Ph.D. in environmental engineering from Duke University. She is an associate professor in the Dept. of Civil and Environmental Engineering at Tulane University. She works in environmental modeling and the area of natural/technological disasters.

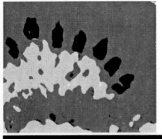

INTERDISCIPLINARY LIVELY APPLICATIONS PROJECT

AUTHORS:
John L. Scharf
jscharf@carroll.edu
(Mathematics and
Engineering)
Carroll College
Helena, MT 59625

Frank Hughes
Johnson Space Center,
NASA
Houston, TX

EDITORS:
Joseph Myers and
Tim Pritchett

CONTENTS

Going into Orbit— Launching the Shuttle

MATHEMATICS CLASSIFICATIONS:
Differential Equations, Numerical Methods, Analytic Geometry

DISCIPLINARY CLASSIFICATIONS:
Physics, Astrophysics

PREREQUISITE SKILLS:
1. Modeling with differential equations
2. Translating physical statements into mathematical form
3. Using a numerical differential equation solver and/or writing a numerical-solver computer code to handle systems of nonlinear ordinary differential equations
4. Using a graphics package

PHYSICAL CONCEPTS EXAMINED:
Classical mechanics, kinetics, and kinematics

COMPUTING REQUIREMENT:
1. Numerical solver for ordinary differential equations
2. Graphing software

UMAP/ILAP Modules 2002–2003: Tools for Teaching, 125–150. Reprinted from *The UMAP Journal* 23 (4) (2002) 429–454. © Copyright 2002, 2003 by COMAP, Inc. All rights reserved.

Contents

Setting the Scene

A launch of the U.S. Space Shuttle.

For several decades, we have been placing objects in orbit around the Earth. The first American to orbit the Earth was astronaut John Glenn. More recently, the Space Shuttle has carried people and cargo into Earth orbits. Satellites, shuttle vehicles, and space stations such as the Russian MIR and the International Space Station all need to be placed in orbit. This project is designed to explore the physics and mathematics that we need to know to get people and objects into orbit around the Earth.

Mechanics is the branch of physics that deals with how objects move under the action of forces. Isaac Newton formulated some simple laws of mechanics, which describe the mechanical interactions of physical bodies. He also formulated a law that describes the gravitational influences that bodies have on one another. Taken together, Newton's Laws of Motion and his Law of Gravitation can begin to help us understand what we need to do in order to put people into orbit.

Part 1. Escape Velocity

"What goes up must come down" is a precept that we accept from an early age; but is it true? Is it possible to throw an object away from the Earth fast enough so that it will never come back?

Requirement 1

According to Newton's Law of Gravitation, the gravitational force that the Earth exerts on an object is proportional to the mass of the object and inversely proportional to the square of its distance from the center of the Earth. This is true provided the object is on or above the surface of the Earth.

Write an expression to describe the gravitational pull of the Earth on an object that is at or above the surface of the Earth. Use the fact that the weight of an object at the surface of the Earth is mg, the mass of the object mass times gravitational acceleration, to determine the constant of proportionality in Newton's Law of Gravitation. Draw a graph that shows how the Earth's gravitational pull on an object varies with its distance from the surface of the Earth.

Requirement 2

For now, you should not worry about the fact that the Earth is rotating about its polar axis or that the atmosphere exerts significant drag forces on an object that we attempt to throw away from the Earth. Newton's Second Law states that the rate of change of the velocity (i.e., the acceleration) of an object of constant mass is proportional to the net force acting on it. The constant of proportionality is the reciprocal of the mass of the object.

Assuming that we throw an object straight up, apply Newton's Second Law to write a differential equation (called the equation of motion) that describes the rate of change of the velocity of the object. To support your equation of motion, use a picture showing the forces that act on the object as it flies away from the Earth.

Requirement 3

Combine your result from **Requirement 2** with the fact that velocity is the time rate of change of position to write a second-order differential equation, or a system of first-order differential equations, to model the motion of the object you are attempting to throw away from the Earth. Also, specify the initial conditions for your model.

Requirement 4

Use a computer system for numerical solutions of ordinary differential equations (ODEs) to solve the equation (or system of equations) from **Requirement 3**. Demonstrate with your solution that, under the assumed conditions, if an object is thrown away from the Earth with a minimum velocity of 36,736 ft/s (or 25,050 mph), it will never return. (The mean radius of the Earth is $R - e = 20.9 \times 10^6$ ft.)

Requirement 5

Discuss how you think the drag of the atmosphere and the rotation of the Earth would change the value for the escape velocity.

Part 2. Orbital Velocity

Suppose for a moment that the Earth is a perfectly smooth sphere (i.e., no mountains or valleys) and that there is no atmosphere. Would it be possible under these conditions to throw a ball horizontally with enough speed so that it would never hit the ground?

The Earth is not flat—it is round. When we are dealing with motions on the Earth that traverse large distances, we cannot ignore this fact in our calculations. Consequently, for these types of problems that involve circular geometry, it is better to use polar coordinates rather than the more familiar xy Cartesian coordinates, which are more suitable for rectangular geometry.

In polar coordinates, a distance from a reference point called the *pole* and an angle from some fixed reference line radiating from the pole specifies position in a plane. **Figure 1** shows point P that is located a distance r from the pole, along a line at an angle θ measured positive counterclockwise from the reference line. The polar coordinates of P are (r, θ).

If point P is moving, then r and θ vary with time. The velocity and acceleration of P can be expressed in terms of r and θ and their time derivatives.

In polar coordinates, there are two components of velocity, one in the radial direction and one in the transverse direction (i.e., perpendicular to the line OP). These two components are given by $v_r = \dot{r}$ and $v_\theta = r\dot{\theta}$. Similarly, the

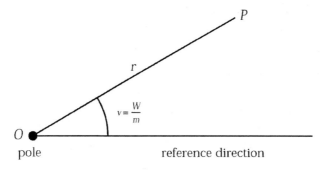

Figure 1. Position of P in polar coordinates.

acceleration of P has a radial and a transverse component. These are given by $a_r = \ddot{r} - r\dot{\theta}^2$ and $a_\theta = r\ddot{\theta} + 2\dot{r}\dot{\theta}$.

Requirement 1

Show that the radial and transverse components of acceleration can be rewritten as

$$a_r = \dot{v}_r \frac{v_\theta^2}{r}, \qquad a_\theta = \dot{\theta}_v + \frac{v_r v_\theta}{r}.$$

Requirement 2

Set up an initial value problem to describe the motion of the ball when it is thrown horizontally with an initial velocity v_0, just slightly above the surface of the perfectly round and perfectly smooth Earth, as shown in **Figure 2**.

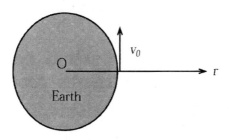

Figure 2. Throwing a ball into orbit.

Use polar coordinates with the pole at the center of the Earth and the reference line through the point on the Earth's surface where the ball is thrown. Assume that the Earth does not rotate and that the only significant force acting on the ball when it is in flight is the Earth's gravitational pull, which is directed toward the center of the Earth. The unknown functions are $r(t)$, $\theta(t)$, $v_r(t)$, and $v_\theta(t)$.

Requirement 3

Using the equations in **Requirement 2**, determine the value of v_0, the initial horizontal velocity of the ball, so that it will orbit the Earth in a circular path just slightly above the Earth's surface. At this speed, how long will it take for the ball to go completely around the Earth?

Requirement 4

Use a computer algebra system to solve the initial value problem and use your solution to plot the orbits of the ball for several initial velocities ranging from the minimum orbital velocity up to and including the escape velocity (see **Part 1, Requirement 4**). What are the shapes of the trajectories of the ball?

Requirement 5

What can you conclude from the results of your investigation in **Requirement 4**?

Requirement 6

Discuss what effect you think the rotation of the Earth would have on the minimum initial throw velocity required to put the ball in orbit.

Requirement 7

Discuss what the effect you think the Earth's atmosphere has on the minimum initial throw velocity required to put the ball in orbit.

Part 3. Let's Get Real

Detailed information about the U.S. Space Shuttle can be found at

```
http://www.ksc.nasa.gov/shuttle/technology/
                          sts-newsref/stsref-toc.html
```

The U.S. Space Shuttle orbiter is launched into Earth orbit with two solid rocket boosters (SRBs) and three main engines (SSMEs). The primary components of the shuttle assembly are the orbiter, two SRBs, three SSMEs and the external fuel tank which provides fuel for the main engines (see **Figure 3**).

When sitting on the launch pad, the total shuttle assembly weighs about 4.5 million lbs. At launch, each SRB has a thrust of 3.3 million lbs. The SSMEs can be throttled over a range of 65% to 109% of their rated power level in 1% increments. A value of 100% corresponds to a thrust level of 375,000 lbs at sea level and 470,000 lbs in a vacuum. A value of 104 percent corresponds to 393,800 lbs at sea level and 488,800 lbs in a vacuum; 109% corresponds to 417,300 lbs at sea level and 513,250 lbs in a vacuum.

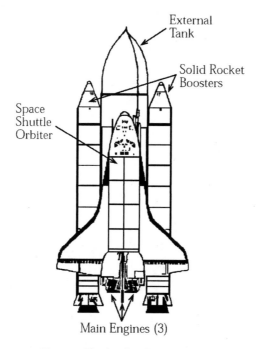

External Tank

Solid Rocket Boosters

Space Shuttle Orbiter

Main Engines (3)

Figure 3. The shuttle vehicle assembly.

The launch of a shuttle begins with main engine ignition and ends approximately 8 min later when the shuttle orbiter enters an orbital trajectory. The ascent occurs in two stages:

- The first stage begins with the ignition of the SRBs and ends two minutes later when the reusable SRBs are jettisoned from the shuttle.

- The second stage begins with SRB separation and ends when the external fuel tank is jettisoned from the shuttle.

The SSMEs are started and brought to 100% power just before the SRBs are ignited to initiate the first stage of the ascent.

At launch, each SRB weighs 1.3 million lbs, which includes 1.1 million lbs of propellant. At 50 s after liftoff, the thrust from the SRBs is reduced to about two-thirds of the maximum. The external fuel tank supplies fuel for the three SSMEs.

At launch, gross weight of the external fuel tank is about 1.66 million lbs, which includes 1.59 million lbs of fuel (liquid oxygen and hydrogen).

The shuttle orbiter travels in near-Earth orbits that range from 115 to 250 mi above mean sea level. The orbiter itself weighs about 230,000 lbs and can carry as many as eight crewmembers. The orbital velocity varies depending upon the altitude of the orbit but is approximately 25,500 ft/s (15,340 mph).

The shuttle systems are very complex and intricate. The model that you will develop and implement in this part of the project attempts to capture some of this complexity; however, the model is still nothing more than a mathematical model and hence many of the intricacies of putting a shuttle into orbit are omitted.

The primary forces acting on the shuttle during ascent to orbit are the thrust from the rocket engines, the gravitational pull of the Earth, and drag due to air resistance. The thrust is used to overcome the weight of the vehicle, lift the shuttle vehicle through the Earth's atmosphere, and accelerate it to the required orbital velocity.

Requirement 1

According to Newton's Second Law (see **Appendix: Newton's Second Law Applied to Rockets**), the net external force on the shuttle, including the thrust from the rockets, is equal to the product of the mass of the shuttle times its acceleration. Consequently, the radial acceleration of the shuttle a_r is equal to the net force in the radial direction divided by the mass of the shuttle assembly; and the transverse acceleration a_θ is proportional to the net force in the transverse direction divided by the mass. For this situation, the mass of the assembly changes during the climb to orbit and the net force includes the thrust from the rocket engines. If the net force in the radial direction is F_r and in the transverse direction it is F_θ, and the mass of the shuttle is m, write two equations that express the relations among a_r, a_θ, F_r, F_θ and m as given by Newton's Second Law.

Requirement 2

As the shuttle climbs into orbit, the thrust from the rocket engines, the aerodynamic forces, and the Earth's gravitational pull all contribute to F_r and F_θ. In addition, the thrust, drag, mass, and weight of the shuttle are all changing during the ascent. Describe qualitatively how each of these quantities (i.e., thrust, drag, weight, and mass) changes and why.

Requirement 3

A diagram that depicts all of the forces acting on an object is called a *free body diagram* of the object. A free body diagram of the space shuttle as it climbs into orbit is shown in **Figure 4.**

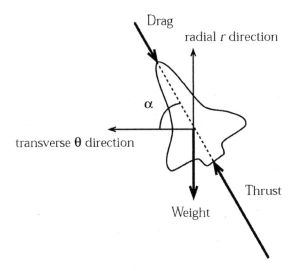

Figure 4. Free body diagram of shuttle during ascent.

The weight is always directed toward the center of the Earth. The thrust due to the engines accelerates the shuttle in the direction it is pointed, and the drag is in the direction opposite the direction of the velocity through the air mass. (We assume that the drag and thrust are collinear and opposite in direction, which is not completely accurate.) If the shuttle is inclined at an angle of α with respect to the transverse (i.e., horizontal) direction, write F_r, the net force in the radial direction, and F_θ, the net force in the transverse direction, each in terms of the drag (D), the weight (W), the thrust (T), and the angle α. Also write the acceleration components a_r and a_θ in terms of these quantities.

The next challenge is to form functions for D, W, T, and m that describe how each of these quantities varies during the ascent to orbit. We start with T, the thrust from the rocket engines.

Requirement 4

Assume that the total thrust from the two solid rocket boosters is 6.6 million lbs at ignition ($t = 0a$) and then it decreases linearly for 50 s to 4.4 million lbs and remains at this level until the SRBs are spent at $t = 120$ s. Write a function that gives the thrust from the SRBs at each instant in time t for all $t \geq 0$. Graph this function for $0 \leq t \leq 150$ s.

Requirement 5

The shuttle main engines can be throttled from 65% to 104%. The total thrust from the three main engines when they are at 100% is about 1.125 million lbs at sea level. Assume for now that the engines operate at sea level. (The decrease in atmospheric pressure as the shuttle climbs through the atmosphere actually results in an increase in the thrust from the engines. This increase will be accounted for later.) For the first 26 s after liftoff, the main engines are at full power, and then they are throttled back to 95% in order to reduce aerodynamic forces as the vehicle passes through the Earth's atmosphere. At $t = 60$ s, the main engines are throttled back up to 100%. (It is approximately at this time that the vehicle is encountering maximum forces due to air resistance.) At $t = 460$ s, the main engines are again throttled back; this time, the thrust is reduced to 65% in order to reduce the g forces on the astronauts and equipment. Main engine cut-off (MECO) occurs at approximately $t = 483.4$ s. Write a function that gives the thrust from the three SSMEs at each instant in time t, for all $t \geq 0$. Graph this function for $0 \leq t \leq 500$ ss.

Requirement 6 (see Part 1, Requirement 1)

According to Newton's Law of Gravitation, the gravitational force that the Earth exerts on an object is proportional to the mass of the object and inversely proportional to the square of its distance from the center of the Earth. This is true provided the object is on or above the surface of the Earth. Write an expression to describe the gravitational pull of the Earth on an object of mass m that is at or above the surface of the Earth, and determine the constant of proportionality. Draw a graph that shows the Earth's gravitational pull on an object as a function of r, the distance of the object from the center of the Earth, for values of $r \geq R_e$, where $R_e = 20.9 \times 10^6$ ft is the mean radius of the Earth.

Requirement 7

In the velocity range of the shuttle, aerodynamic forces due to the space vehicle's motion through the Earth's atmosphere during ascent to orbit are quite large. The air resistance force on the vehicle is proportional to the density of the atmosphere, the geometry and orientation of the spacecraft relative to its direction of motion through the air, and the square of the speed of the vehicle relative to the air mass. Using the two polar components of velocity, write an expression for the aerodynamic drag force on the shuttle system as it climbs to orbit. Use ρ to represent the air density and k to represent the geometry/orientation factor.

Requirement 8

The density of the Earth's atmosphere varies with height above the Earth and is given by $\rho = \rho_0 e^{-h/28276}$, where $\rho_0 = 2.33 \times 10^{-3}$ slugs/ft^3 is the air density at sea level and h is the distance above the surface of the Earth in feet. (Note: $h = r - R_e$.) Assuming that $k = 430$ lb-ft-s^2/slug, rewrite your function for air resistance in **Requirement 7** to include the variation in air density with altitude. Sketch a qualitative graph that you think would depict the air resistance force on the shuttle as a function of h, during the climb to orbit.

Requirement 9 (see Part 2, Requirement 1)

Recall that in polar coordinates, there are two components of velocity, one in the radial direction and one in the transverse direction. These two components are given by $v_r = r$ and $v_\theta = r\dot\theta$. Similarly, acceleration has a radial and a transverse component. These are given by $a_r = \ddot{r} - r\dot\theta^2$ and $a_\theta = r\ddot{theta} + 2\dot{r}\dot\theta$. Show that the components of acceleration can be rewritten as

$$a_r = \dot{v}_r \frac{v_\theta^2}{r}, \qquad a_\theta = \dot\theta_v + \frac{v_r v_\theta}{r}.$$

Requirement 10

During the ascent, the mass of the shuttle assembly decreases dramatically. This happens for two reasons:

- First, the shuttle engines burn fuels and throw off mass as exhaust gases.

- Second, the SRB's and the external tank are jettisoned once they are spent.

The rate at which the shuttle engines consume mass as fuel is determined by the design of the engines. We assume that the rate of mass reduction is proportional to the thrust delivered by the engines. Assume that the SRBs consume fuel at the rate of 105.4 slugs/s per million lbs of thrust and that the SSMEs consume 94.7 slugs/s per million lbs of thrust. The spent SRBs each have a mass of 6250 slugs and they are jettisoned at $t = 120$ s. The empty external tank has a mass of 2062 slugs and it is jettisoned at $t = 480$ s. Assume that it takes 1 s to jettison the two SRBs and 1 s to jettison the external tank. Write a differential equation that expresses the rate of change of mass of the shuttle system in terms of the thrust from the SRBs and SSMEs. Also include in your function the change in mass that occurs when the SRBs and the external tank are jettisoned.

Requirement 11

The thrust from the SRBs and the SSMEs is about 25% higher in a vacuum than it is at sea level. Determine a factor that would multiply the thrust functions (see **Requirements 4** and **5** in this part) to increase the thrust with altitude. (Hint: In **Requirement 8** of this part, there is a function that accounts for the decrease in atmospheric density as altitude increases.)

Requirement 12

To change the direction of flight during the ascent, the shuttle uses thrust vector control. The direction of the thrust vector (see **Figure 4**) is controlled to gradually change the direction of flight as the shuttle ascends to orbit. Specifying the angle α as a function of time does this. (In reality, α is adjusted continuously during the ascent to keep the spacecraft on a predetermined trajectory that will bring it into the desired orbit; however, we will specify, a priori, how α will vary with time during the ascent.) The shuttle goes straight up ($\alpha = 90°$) for 6 s until it clears the lightning mast on the launch pad. Over the next 20 s, it pitches to $\alpha = 78°$. (During the same time interval, it rolls so that the shuttle orbiter is under the external tank.) Between $t = 26$ s and $t = 240$ s, the shuttle pitches at a uniform rate to $\alpha = 21°$. After $t = 240$ s, the thrust is in the direction of the velocity vector. Write an expression for $\alpha(t)$.

Requirement 13

A system of differential equations that model the shuttle launch to orbit can now be formulated. The unknown functions are $r(t)$, $\theta(t)$, $v_r(t)$, $v_\theta(t)$, and $m(t)$. Write the five first-order differential equations that model the shuttle launch.

Requirement 14

Before solving the system, it is necessary to specify the initial conditions (i.e., the values of the unknown functions at $t = 0$ s). Assume that the launch will be due east from Cape Canaveral. Because of the Earth's rotation, the shuttle's initial transverse velocity is 1530 ft/s due east. Write the initial values for each of the unknown functions.

Requirement 15

Use a numerical differential equation solver to solve the system of equations that models the shuttle launch. Plot the solutions as functions of time. How high is the orbit? How long does it take for the shuttle to complete one orbit? What is the orbital velocity?

Requirement 16

Generate a parametric plot that shows the Earth and the trajectory of the shuttle from launch through one complete orbit.

Requirement 17

Explore what happens when the length of the main engine burn is extended to $t = 485$ s and when it is reduced to $t = 482$ s.

Appendix: Newton's Second Law Applied to Rockets

For a rocket, where the mass of the rocket changes as it ejects fuel through the engine nozzles, Newton' Second Law must be applied in its general form. Specifically, it applies to the entire system of mass including the rocket and the fuel—the fuel that is retained as well as the fuel that is ejected. In this situation, Newton's Second Law states that the net force is equal to the time rate of change of the total linear momentum of the system, that is $\vec{F}_{external} = d\vec{P}/dt$, where $\vec{F}_{external}$ is the resultant external force on the system and $\vec{P}$ is the total linear momentum.

To see how this works, we must consider the momentum of the rocket together with the momentum of the retained fuel as well as the momentum of the ejected fuel. Consider what happens in a short interval of time Δt. At the beginning of the time interval, the mass of the rocket plus the fuel that will be ejected is $m_\Delta m$, and velocity of the rocket and fuel is $\vec{v}$, as shown in **Figure A.1**.

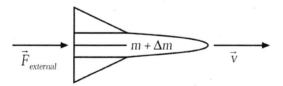

Figure A.1. Rocket and fuel before fuel ejection.

At the end of the time interval Δt, the ejected fuel of mass Δm has a velocity $\vec{u}$, while the velocity of the rocket changes by an amount $\Delta \vec{v}$. The configuration of the system at the end of the time interval θ is as shown in **Figure A.2**.

The change in linear momentum during Δt is

$$\Delta \vec{P} = \left[m(\vec{v} + \Delta\vec{v}) + (\Delta m)\vec{u} \right] - (m + \Delta m)\vec{v} = m\Delta\vec{v} + \Delta m(\vec{u} - \vec{v}),$$

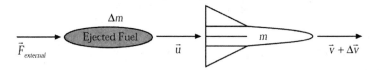

Figure A.2. Rocket and fuel after fuel ejection.

and the average rate of change of momentum is

$$\frac{\Delta \vec{P}}{\Delta t} = m \frac{\Delta \vec{v}}{\Delta t} + \frac{\Delta m}{\Delta t} (\vec{u} - \vec{v}).$$

In the limit, as $\Delta t \to 0$, the instantaneous rate of change of momentum is

$$\frac{\vec{P}}{dt} = m \frac{\vec{v}}{dt} + \frac{dm}{dt} (\vec{u} - \vec{v}) \qquad |textor \qquad \frac{\vec{P}}{dt} = m\vec{a} + \frac{dm}{dt} (\vec{u} - \vec{v}),$$

where $\vec{a}$ is the acceleration of the rocket, dm/dt is the rate at which the rocket engines eject fuel mass from the rocket, and $\vec{u} - \vec{v}$ is the velocity of the ejected fuel relative to the rocket. Therefore, for the rocket-fuel system, the general form of Newton's Second Law is

$$\vec{F}_{\text{external}} = m\vec{a} + \frac{dm}{dt} (\vec{u} - \vec{v}) \qquad \text{or} \qquad \vec{F}_{\text{external}} + \frac{dm}{dt} (\vec{v} - \vec{u}) = m\vec{a}.$$

The term $\frac{dm}{dt}(\vec{v} - \vec{u})$ is the thrust on the rocket from the ejected fuel. Its magnitude depends on the dm/dt, the rate at which the engines eject mass from the rocket, and $\vec{v} - \vec{u}$, the speed of the ejected fuel relative to the rocket. If we let $\vec{T} = \frac{dm}{dt}(\vec{v} - \vec{u})$, then Newton's Second Law for the rocket is

$$\vec{F}_{\text{net}} = \vec{F}_{\text{external}} + \vec{T} = m\vec{a} \qquad \text{or} \qquad \vec{F}_{\text{net}} = m\vec{a},$$

where it is now understood that m varies with time and $\vec{F}_{\text{net}}$ is the net external force on the rocket including the thrust from the fuel that is ejected through the engines.

References

Hibbeler, R.C. 1995. *Engineering Mechanics: Statics and Dynamics.* 7th ed. Englewood Cliffs, NJ: Prentice-Hall.

National Aeronautics and Space Administration. `http://www.ksc.nasa.gov/shuttle/technology/sts-newsref/stsref-toc.html` .

Serway, R.A. 1996. *Physics for Scientists and Engineers.* 4th ed. Philadelphia, PA: Saunders College Publishing.

Thomas, G.B., Jr., and Ross L. Finney. 1996. *Calculus and Analytic Geometry.* 9th ed. Reading, MA: Addison Wesley.

Sample Solutions

Part 1. Escape Velocity

Requirement 1

If W is the gravitational pull of the Earth then W is proportional to m/r^2 , or $W = km/r^2$. At or near the surface of the Earth, $r = R_e$, the radius of the Earth, and $W = mg$, where g is the gravitational acceleration, giving $k = gR_e^2$. Substituting this value for k back into the gravitation law gives $W = mg(R_e^2/r^2)$. If h is the altitude of an object above the surface of the Earth (i.e., above mean sea level), then $r = R_e + h$ and Newton's Law of Gravitation can be written as

$$W = mg\,\frac{R_e^2}{(R_e + h)^2},$$

which can be rewritten in nondimensional form as

$$\frac{W}{mg} = \frac{1}{\left(1 + \dfrac{h}{R_e}\right)^2}.$$

Figure S.1 shows a graph of W/mg versus h/R_e.

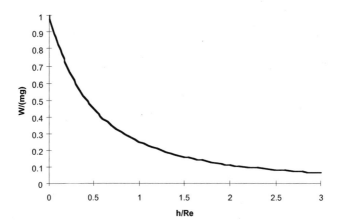

Figure S.1. Weight vs. altitude.

Requirement 2

A free body diagram of an object as it flies straight up from the Earth is shown in **Figure S.2**.

Taking the direction away from the Earth to be positive, the acceleration is then given by $a = -W/m$ or $\dot{v} = -W/m$.

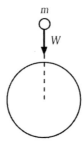

Figure S.2. Free body diagram of object flying away from Earth.

Requirement 3

Since $\dot{r} = v$ and $W = mg(R_e^2/r^2)$], the differential equation in **Requirement 3** can be rewritten as $\ddot{r} = -g(R_e^2/r^2)$. The initial conditions for this second-order equation are $r(0) = R_e$ and $\dot{r}(0) = v_0 a$, the initial upward velocity of the object being thrown. To rewrite the equations in terms of altitude, replace r with $R_e + h$ to obtain

$$\ddot{h} = -g\,\frac{R_e^2}{(R_e + h)^2}$$

with initial conditions $h(0) = 0$ and $\dot{h}(0) = v$.

An alternative is to use the following equivalent system of first-order differential equations to model the situation: $\dot{v} = -g(R_e^2/r^2)$ and $\dot{r} = v$. The initial conditions for this system are $r(0) = R_e$ and $v(0) = v_0$.

Requirement 4

To solve the system numerically, values for g, R_e , and v_0 must be specified. Using Mathematica to solve the second-order equations in h (see **Requirement 3**) gives solutions that are summarized by the graphs in **Figure S.3**.

By trial and error, the escape velocity is estimated to be about 36,750 ft/s. The graphs of the solutions for this case are in **Figure S.3c**. The time interval can be extended to show that the object continues moving away.

Requirement 5

The atmosphere will tend to slow an object as it passes through, because of aerodynamic drag forces. Consequently, the object would have to be thrown at a higher velocity to escape. On the other hand, because the Earth is rotating, the object has an initial velocity before it is thrown, provided it is not being thrown from one of the rotational poles. Consequently, the object could be thrown at a smaller velocity and still escape. The closer to the equator, the greater the objects velocity due to the Earth's rotation, and the lesser the throw velocity that is required to escape.

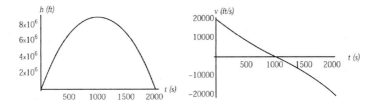

Figure S.3a. Altitude and speed for $v_0 = 20,000$ ft/s.

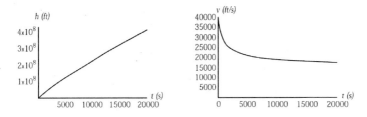

Figure S.3b. Altitude and speed for $v_0 = 40,000$ ft/s.

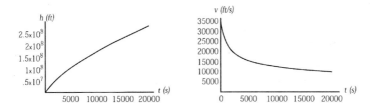

Figure S.3c. Altitude and speed for $v_0 = 36,750$ ft/s.

Part 2. Orbital Velocity

Requirement 1

First, the radial acceleration is $a_r = \ddot{r} - r\dot{\theta}^2 = \ddot{r} - (r\dot{\theta})^2/r$. However, since $\ddot{r} = \dot{v}_r$ and $r\dot{\theta} = v_\theta$, the radial acceleration can be written as $a_r = \dot{v}_r - v_\theta^2/r$. The key to transforming the transverse acceleration is to realize that $\dot{v}_\theta = \frac{d}{dt}(r\dot{\theta}) = \dot{r}\dot{\theta} + r\ddot{\theta}$ and to rewrite a_θ as

$$a_\theta = r\ddot{\theta} + 2\dot{r}\dot{\theta} = \dot{r}\dot{\theta} + r\ddot{\theta} + \frac{\dot{r}(r\dot{\theta})}{r} = \dot{v}_\theta + \frac{v_r v_\theta}{r}.$$

Requirement 2

While the object is in flight, the only force on it is the Earth's gravitational pull. Consequently, $F_r = -mg(R - e/r)^2$ and $F_\theta = 0$. The force F_r is negative because the positive radial direction is away from the Earth and the pull of gravity is toward the Earth. Newton's Second Law requires that $F_r = ma_r$ and

$F_\theta = ma_\theta$. Substituting the expressions for the components of acceleration and force gives

$$-mg\left(\frac{R_e}{r}\right)^2 = m\left(\dot{v}_r - \frac{v_\theta^2}{r}\right) \quad \text{and} \quad 0 = m\left(\ddot{v}_\theta + \frac{v_r v_\theta}{r}\right).$$

Solving for the derivatives gives four autonomous nonlinear first-order equations:

$$\dot{v}_r = \frac{v_\theta^2}{r} - g\left(\frac{R_e}{r}\right)^2, \quad \dot{v}_\theta = -\frac{v_r v_\theta}{r}, \quad \dot{r} = v_r, \quad \text{and} \quad \dot{\theta} = \frac{v_\theta}{r},$$

with the initial conditions $v_r(0) = 0$, $v_\theta(0) = 0$, $r(0) = R_e$, and $\theta(0) = 0$.

Requirement 3

If the ball is to travel in a circular orbit just at the Earth's surface, then $r = R_e$ and $\dot{v}_r = v_r = 0$ for all values of t. Substituting these into the first two differential equations in **Requirement 2** gives $v_\theta = \sqrt{gR_e} = 25{,}861$ ft/s = 17,633 mph. The circumference of the Earth is about 131.32×10^6 ft $\approx 24{,}871$ mi. At the orbital speed of 17,633 mph, it would take the ball about 1 h 25 min to complete one orbit.

Requirement 4

Using the NDSolve[] command in Mathematica, you obtain numerical solutions and the orbits for various initial velocities, as shown in the parametric graphs of **Figure S.4**.

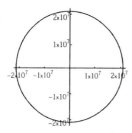
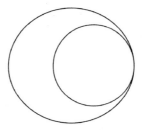

Figure S.4a. $v_0 = 25{,}861$ ft/s: The orbit and the surface of the Earth overlap.

Figure S.4b. $v_0 = 30{,}000$ ft/s: The orbit is an ellipse.

Figure S.4c. $v_0 = 36{,}740$ ft/s (escape velocity): The orbit is a parabola and the ball would never return.

Requirement 5

For an initial velocity of 25,861 ft/s, the orbit will be a circle just at the Earth's surface. For initial velocities between 25,861 ft/s and 36,740 ft/s, the orbits are elliptical with the ball eventually returning to the place where it was thrown. For velocities greater than (or equal to) 36,740 ft/s, the ball escapes the Earth's gravitational pull along a hyperbolic (parabolic) trajectory.

Requirement 6

Other than at the rotational poles of the Earth, the rotation of the Earth will reduce the initial throw velocity required to put the ball in orbit. The closer to the equator, the less the initial throw velocity will have to be. This assumes that the throw is in the same direction as the Earth's rotation. This also explains why launch centers are typically located so that they have water or uninhabited areas to their east.

Requirement 7

As the ball flies through the Earth's atmosphere, there would be an air resistance force which would slow the ball and cause it to fall eventually to Earth, no matter how fast you could throw it. In order to reduce this force, it is necessary to get the object above the dense layer of the atmosphere that is relatively close to the Earth's surface and then accelerate the object laterally to the required orbit velocity.

Part 3. Let's Get Real

Requirement 1 (See Part 2, Requirement 2)

Newton's Second Law requires that $a_r = F_r/m$ and $a_\theta = F_\theta/m$. The mass m varies with time, and the components F_r and F_θ of the resultant force on the shuttle include the components of the rocket thrust (see the **Appendix**).

Requirement 2

The thrust from the engines changes as the engines are throttled; however, the most significant changes occur when the solid rocket boosters burn out and are jettisoned, and at main engine cut-off (MECO), when the shuttle main engines are shut off just prior to orbit insertion. The aerodynamic drag force is zero initially, but as the shuttle accelerates from the launch pad and velocity accumulates the drag forces increase. Quickly, as the shuttle climbs out of the Earth's atmosphere, the drag forces decay to near zero. The gravitational pull of the Earth decreases during the ascent because the shuttle moves away from the Earth. It is important to realize, however, that while the gravity force decreases, it does not decrease substantially. In the shuttle's low Earth orbit, the gravitational pull of the Earth on the shuttle is no less than 90% of that on the surface of the Earth. The mass of the shuttle changes for two reasons: the rockets throw off huge amounts of burned fuel, and pieces (the SRBs and external tank) are jettisoned when they are no longer needed.

Requirement 3

If T is the thrust, D is the drag, and W is the Earth's gravitational pull, then the net force component in the positive radial direction (i.e., away from the Earth) is $F_r = (T - D)\sin\alpha - W$. The net force in the transverse direction is $F_\theta = (T_D)\cos\alpha$. The corresponding accelerations are

$$a_r = \frac{1}{m}\left[(T - D)\sin\alpha - W\right] \quad \text{and} \quad a_\theta = \frac{1}{m}\left[(T - D)\cos\alpha\right].$$

Requirement 4

The thrust from the solid rocket boosters (in millions of lbs of force) is

$$T_{\text{SRB}} = \left(6.6 - 2.2\,\frac{t}{50}\right)\left[\mu(t) - \mu(t - 50) - \mu(t - 120)\right],$$

where μ is the unit step function. The graph of this function is shown in **Figure S.5**.

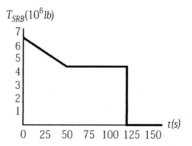

Figure S.5. Thrust schedule for the solid rocket boosters.

Requirement 5

The thrust from the main engines (in millions of lbs of force) is

$$T_{\text{ME}}(t) = 1.125\left[\mu(t) - mu(t - 26) - \mu(t - 60)\right] + 1.125\left[\mu(t - 60) - \mu(t - 460)\right]$$
$$+ 0.731\left[\mu(t - 460) - mu(t - 480)\right].$$

The graph of this function is shown in **Figure S.6**.

Requirement 6 (See Part 1)

Newton's Law of Gravitation is $W = mg(R_e/r)^2$. The graph of W/mg vs. r/R_e is shown in **Figure S.7**.

For a shuttle orbit that is 200 mi above the surface of the Earth, $r/R_e = 1.05$ and $W/mg = 0.91$. That is, the Earth's gravitational pull on the shuttle in orbit is about 91% of what it is on the surface of the Earth.

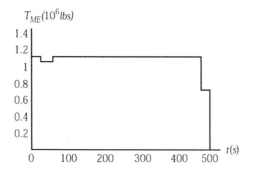

Figure S.6. Thrust schedule for the shuttle main engines.

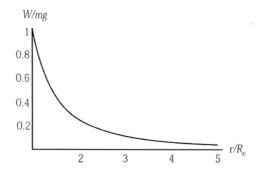

Figure S.7. Gravitational force vs. distance from center of Earth.

Requirement 7

The aerodynamic drag force on the shuttle is

$$D = k\rho\big[v_r^2 + (v_\theta - v_{\theta0})^2\big].$$

The quantity $v_{\theta0}$ is the initial transverse velocity of the shuttle due to the Earth's rotation. We assume that the air mass at the launch pad is moving with the shuttle and the Earth (i.e., the winds are calm). Therefore, to calculate the speed of the shuttle relative to the air mass through which it is passing during the ascent, the component of the velocity due to the Earth's rotation must be subtracted from the transverse component v_θ of the velocity.

Requirement 8

Accounting for the variation in air density with altitude, and given the value for the drag coefficient, the air resistance force on the shuttle is

$$D = 430 \left(2.33 \times 10^{-3}\right) e^{\frac{r - R_e}{28276}} \left(v_r^2 + (v_\theta - v_{\theta0})^2\right)$$
$$= 1.00e^{\frac{r - R_e}{28276}} \left(v_r^2 + (v_\theta - v_{\theta0})^2\right).$$

Initially, the air resistance force is zero; but as the shuttle's speed increases, the air resistance force will also increase. As the shuttle goes up, the density of the air through which it is passing decreases to near zero. A qualitative graph of the drag force is shown in **Figure S.98**.

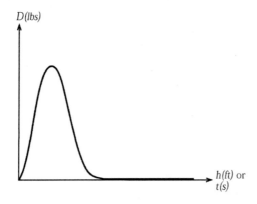

Figure S.8. Qualitative graph of air resistance force vs. time or altitude.

Requirement 9

See **Part 2, Requirement 1**.

Requirement 10

The differential equation governing the rate of change of change of mass of the system during ascent to orbit is

$$\frac{d}{m} = -117.9T_{\text{ME}}(t) - 94.77T_{\text{SRB}}(t) - 12500\big[\mu(t-119.5) - \mu(t-120.5)\big]$$
$$- 2062\big[\mu(t-479.5) - \mu(t-480.5)\big].$$

The first two terms account for the changing mass because fuel is being burned and ejected by the solid rocket boosters and by the main engines, the third term is related to jettisoning of the SRBs, and the last term accounts for jettisoning of the external tank.

Requirement 11

As the shuttle climbs, atmospheric pressure decreases. resulting in improved performance of the shuttle engines. Since performance improves with altitude, the air density reduction factor should apply. The factor to multiply the thrust from the engines is

$$1.25 - 0.25e^{\frac{r-R_e}{28276}}.$$

At the surface of the Earth, the thrust factor is 1; and as the shuttle climbs out of the atmosphere it increases to 1.25.

Requirement 12

The following is a possible thrust control schedule (where α is in degrees):

$$\alpha(t) = 90\big[\mu(t) - \mu(t-6)\big] + \left(90 - 12\,\frac{t-6}{20}\right)\big[\mu(t-6) - \mu(t-26)\big]$$

$$+ \left(78 - 57\,\frac{t-26}{214}\right)\big[\mu(t-26) - \mu(t-240)\big] + \tan^{-1}\left(\frac{v_r}{V_\theta}\right)\mu(t-240).$$

Requirement 13

The five governing differential equations are:

$$\dot{v}_r = \frac{v_\theta^2}{r} - g\left(\frac{R_e}{r}\right)^2 + \frac{1}{m}\left[(T_{\text{SRB}} + T_{\text{ME}})\left(1.25 - 0.25e^{\frac{r-R_e}{28276}}\right) - D\right]\sin\alpha,$$

$$\dot{v}_\theta = \frac{v_r v_\theta}{r} + \frac{1}{m}\left[(T_{\text{SRB}} + T_{\text{ME}})\left(1.25 - 0.25e^{\frac{r-R_e}{28276}}\right) - D\right]\cos\alpha,$$

$$\dot{m} = -117.9T_{\text{ME}} - 94.77T_{\text{SRB}} - 12500\big[\mu(t-119.5) - \mu(t-120.5)\big]$$
$$- 2062\big[\mu(t-479.5) - \mu(t-480.5)\big],$$

$$\dot{r} = v_r, \qquad \dot{\theta} = \frac{v_\theta}{r}.$$

Requirement 14

The five initial conditions are

$$v_r(0) = 0, \quad v_\theta(0) = 1530, \quad m(0) = \frac{4.5 \times 10^6}{g}, \quad r(0) = R_e, \quad \theta(0) = 0.$$

Requirement 15

Using the NDSolve[] command in Mathematica, you can obtain numerical solutions for $r(t)$, $\theta(t)$, $v_r(t)$, $v_\theta(t)$, and $m(t)$. In **Figure S.9** are two sets of graphs for these functions. The left-hand set is over the time period for two complete orbits (about 3 h) and the right-hand is over the first 12 min after liftoff. The orbit is 174.4 mi above the surface of the Earth, the orbital velocity is 25,803 ft/s, and it takes 1 h 30 min to complete one orbit.

Requirement 16

Figure S.10 shows the shuttle orbit to scale.

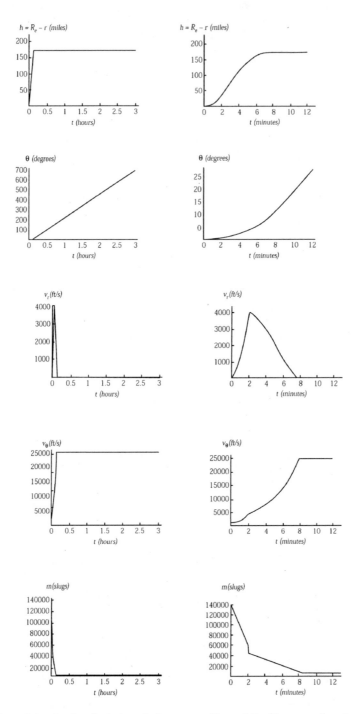

Figure S.9a. Graphs of solutions during the first two complete orbits (about 3 h).

Figure S.9b. Close-ups of graphs of solutions during the first 12 min after liftoff.

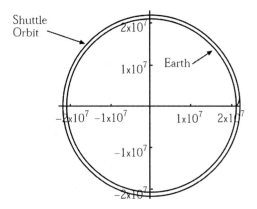

Figure S.10. The shuttle orbit, to scale.

Requirement 17

With the main engine burn extended to 485 s, the orbit swings out wider on the side of the Earth away from the launch site, as shown in **Figure S.11a**. With the main engine burn reduced to 482 s, the trajectory swings in closer on the side of the Earth away from the launch site, as shown in **Figure S.11b**.

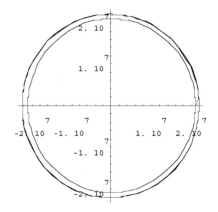

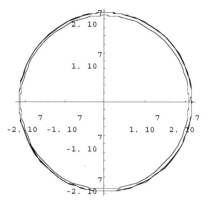

Figure S.11a. The shuttle orbit with main engine burn extended to 485 s; the orbit swings wider at left of the figure, the side of the Earth away from the launch site (at right of figure).

Figure S.11b. The shuttle orbit with main engine burn reduced to 482 s; the orbit swings in closer at left of the figure, the side of the Earth away from the launch site (at right of figure).

Title: Going into Orbit—Launching the Shuttle

Notes for the Instructor

This ILAP is designed for an introductory course in differential equations that includes nonlinear systems. It would also be appropriate for an introductory mechanics course in physics or engineering dynamics.

Students doing this project need to have access to, and know how to use, a numerical ODE solver and graphing software. In addition, they must be able to apply Newton's Laws of Motion and Gravitation to generate the equations of motion for the shuttle, and they must be able to translate physical statements into a mathematical form.

Parts 1 and 2 (Escape Velocity and Orbital Velocity) are limited in scope and can be used without Part 3 for a more focused project. This more focused project can be done in a one- to two-hour time period in a computer lab, or you can assign it to be completed outside of class. Part 3 is much more substantial than Parts 1 and 2 and should be assigned as a major project for a student, or group of students, to complete outside of class.

Part 1 (Escape Velocity) is a relatively straightforward application of Newton's Laws to describe motion along a straight line. Polar coordinates are used in Part 2 (Orbital Velocity) to set up the equations for orbital motion in two dimensions. Parts 1 and 2 are designed to illustrate what needs to be done in order to put the shuttle in orbit, setting a basis for Part 3, Let's Get Real. Part 3 requires an elaborate model that includes mass change of the system, air resistance, variable thrust, and realistic initial conditions. It leads to a system of five nonlinear differential equations.

An extension of Part 1 (Escape Velocity) and Part II (Orbital Velocity) that you may wish to include in a physics or an engineering mechanics course is to have students use energy methods in addition to the direct integration of the equations of motion. An extension to Part 3 is to have students investigate orbit strategies that are used to move from one orbit to another (i.e., the Hohman transfer) and to implement these using their model. This extension can lead to an exploration of the methods that are used for rendezvous in orbit. For example, if the shuttle and the space station are in the same orbit but the shuttle is behind the station, how could the shuttle catch up to the station for a rendezvous?

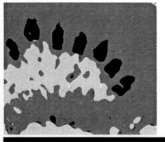

I N T E R D I S C I P L I N A R Y   L I V E L Y   A P P L I C A T I O N S   P R O J E C T

AUTHORS:
George Ashline
(Mathematics)
gashline@smcvt.edu

Alain Brizard
(Chemistry and Physics)
abrizard@smcvt.edu

Joanna Ellis-Monaghan
(Mathematics)
jellis-monaghan@smcvt.edu

St. Michael's College
Colchester, VT 05439

EDITOR:
Paul J. Campbell
Beloit College, WI

CONTENTS

Water Rockets in Flight: Calculus in Action

MATHEMATICS CLASSIFICATIONS:
   Calculus

DISCIPLINARY CLASSIFICATIONS:
   Physics (mechanics)

PREREQUISITE SKILLS:
   1. Trigonometry
   2. Vector calculus for three-dimensional space curves
   3. Single-variable calculus for the simpler height vs. time
      model.

PHYSICAL CONCEPTS EXAMINED:
   1. Orthogonality of speed and acceleration in horizontal
      and vertical directions
   2. Relationship between position, velocity, and acceleration
   3. Wind resistance

COMPUTING REQUIREMENT:
   Computer algebra system or graphing calculator, for curve
   fitting

Contents

Abstract

We describe an easy and fun experiment using water rockets to demonstrate some of the concepts of multivariable calculus. After using video stills from a single water rocket launch to generate the raw data, we develop a model to analyze the rocket flight.

Because of factors such as rocket propulsion and wind effects, the water rocket does not follow the parabolic projectile trajectory commonly found in textbooks. Instead, we use polynomial interpolation to calculate the X, Y, and Z coordinate functions of the rocket as a function of time during its entire flight. We then use methods from multivariable calculus to analyze the flight and to estimate quantities such as the maximum height reached by the rocket and curvature of the flight path that are not apparent from direct observation. Examination of first and second time derivatives of the rocket coordinates allows us to identify the thrusting, coasting, and recovery stages of the rocket flight, and comparison to the parabolic model shows the effects of the wind.

We offer two variations of the Module:

- One is very similar to that described above but uses a least-squares fit instead of polynomial interpolation to determine the coordinate functions.

- The other is a simpler model based on a one-variable polynomial fit giving the height of the rocket as a function of time, suitable for a first-semester calculus course.

Concluding resources include an annotated overview of Internet sites, references for curve-fitting techniques using linear algebra, an auxiliary set of video stills, and Maple 8 code for generating the results.

1.   Setting the Scene

Water rockets are cheap, reusable, easy to launch, and have a very high fun-to-nuisance ratio. They also provide a simple example of some of the fundamental aspects of a model rocket flight. Because of factors such as rocket propulsion and wind effects (which can be classified as systematic if the wind is steady or as random if the wind is gusty), their flight paths are more complex than those of simple projectiles.

The goal of this Module is to model the flight path of a single water rocket launch and then use the tools of calculus to analyze the rocket's performance. Most calculus textbooks include both one- and two-dimensional parabolic models describing the flight of a projectile. In the one-dimensional case, the function $s(t) = vt - \frac{1}{2}gt^2$ models the vertical position of a projectile with respect to time t, where g (= 9.81 m/s^2) is the gravitational constant and v is the initial velocity. In the two-dimensional case, the vector-valued function $r(t) = \langle (S\cos\theta)t, (S\sin\theta)t - \frac{1}{2}gt^2 \rangle$ models the planar trajectory of a projectile, where S is the initial speed and θ is the initial launch angle as measured from the ground. In both cases, it is assumed that, other than the initial boost acceleration, gravity is the only force acting on the rocket (e.g., air resistance and Coriolis effects associated with the rotation of the Earth are ignored).

We originally wanted a simple and engaging experiment that would provide the raw data for using these models. However, we did not have ready access to a large windless space (such as a hangar) nor a mechanism to measure the initial velocity needed to illustrate these standard models.

Rather than trying to fit the experimental data to a parabolic model, we chose instead to generate and analyze a single data set and explore the vagaries introduced by wind and variable thrust. We then needed a suitable projectile. On the one hand, tossing a ball vertically into the air did not seem exciting enough. On the other hand, anything involving rocket propulsion by fuel combustion seemed a little too exciting and hence logistically too difficult; also, they move too fast for easy measurements and are too expensive. We wanted something cheap, easy, and safe. Water rockets are the perfect solution. Furthermore, their behavior is more varied than that of projectiles modeled by the parabolic functions. They have just enough complexity in their flight paths to provide an opportunity to put the skills and concepts learned in calculus to work in a substantial way.

A water rocket consists of a tapered plastic chamber about 13 cm long with small fins and a little hole in the base. The chamber is partially filled with water and then air is forced into the chamber with a manual air pump that clamps onto the base. This clamp is also the launch mechanism. See **Figure 1**.

When the rocket is released, the air and water escapes rapidly through the small hole in the base of the rocket, providing the power for the first stage of the rocket's flight, the *thrusting* (or boost) stage. The rocket then continues to soar upwards, although more and more slowly, with now its acceleration only affected by gravity, air resistance, and (possibly) wind; this second stage of the rocket's flight is the *coasting* stage. The rocket then returns to the ground in the *recovery* stage (defined as the part of the rocket flight path from the time it reaches its maximum height until it lands). This terminology is borrowed from more sophisticated rockets that often have a recovery mechanism (such

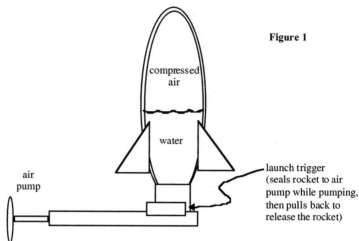

Figure 1

compressed
air

water

air
pump

launch trigger
(seals rocket to air
pump while pumping,
then pulls back to
release the rocket)

Figure 1. Schematic diagram of a water rocket.

as a parachute) to minimize damage to the rocket upon landing. Water rockets, however, do not need parachutes, since they are quite sturdy; and as long as they land on a soft surface such as turf, they will be undamaged.

Water rockets are quite lightweight, hence even a slight crosswind can significantly affect their flight paths. Indeed, only crosswind effects can cause the path of the rocket to exhibit nonplanar features, since gravity, thrust, and air resistance are all planar forces.

Because of the additional forces acting on a water rocket, simple projectile models are inadequate for analyzing its flight path. Excellent recent papers give well-developed models for the flight path of a generic water rocket; however, Prusa [2000] is beyond the scope of a typical undergraduate calculus sequence, and Finney [2000] considers only height vs. time rather than the three-dimensional space curve that we wanted to consider. Furthermore, since these models require ideal launch conditions (e.g., windless conditions and perfectly vertical launches) that are very difficult to achieve in practice, we chose instead to analyze a *specific* rocket-flight data set simply by fitting a smooth curve to a finite number of data points along the flight path.

To get the raw data, we videotaped the flight of a water rocket, with a building of known height in the background. On the day of the launch, there was gusting wind of perhaps 8 to 24 km/h (5 to 15 mph); so naturally the rocket was blown off its planar (and parabolic) course. We noted the position of the rocket against the building in the video stills and then used building blueprints to measure the rocket's horizontal and vertical positions with respect to the building. With basic trigonometry and the ground distances, we converted this information into estimates for the three-dimensional position of the rocket. Having found these coordinates, we then applied the curve-fitting capabilities of the computer algebra system Maple to construct a model of the flight path.

From the vantage point of a single video camera position, information concerning the depth position of the rocket is not available; so nonplanar crosswind effects are not directly observable through the present analysis. Since we expect wind gusts generically to possess planar components, wind effects are expected to be characterized by segments of the flight path with zero curvature (i.e., with parallel velocity and acceleration vectors).

With this model, we can use the tools of calculus to answer questions that could not be addressed just by watching the launch. For example, how high did the rocket go? How far did it travel? How sharp was its turn around? How long did the thrusting stage last? To what extent did wind effects modify the action of gravity? How far was the rocket blown off course by the wind? Answering these questions requires computing and analyzing the derivatives of the spatial coordinates (i.e., analyzing the velocity and acceleration vectors) and finding the Frenet-Serret curvature of the flight path (here, our model explicitly assumes a zero-torsion flight path). Although a planar parabolic-path model does not provide a good approximation of the rocket flight path, a comparison to our curve gives a good sense of the effects of the wind and thrust on the behavior of the rocket.

We use a sixth-degree polynomial in our model. For modeling the height of a rocket flight, an appropriate polynomial is an even-degree one that is initially increasing and ultimately decreasing. We found that second- and fourth-degree polynomials did not capture the wind effects and did not fit the data points well. Polynomials of degree greater than six fit the data points quite well (as might be expected) but were poor models for the flight; they varied too much laterally and "flattened out" at the top. Hence, they could not be used to get good estimates of the maximum height.

In the absence of air resistance and wind effects, there is no lateral acceleration, so the $X(t)$ and $Y(t)$ coordinate functions are expected to be linear in time t. However, we did observe wind effects and hence we use sixth-degree polynomials for these coordinate functions too, to capture this phenomenon.

Experimentation with different curves can be quite instructive though, and we highly recommend doing it to see the effects of the various parameters on a model.

While most of this paper involves a three-dimensional space curve created with polynomial interpolation, we also include two possible ways of modifying the Module. One uses the least-squares fitting method instead of polynomial interpolation to fit the coordinate functions. Doing so gives slightly different answers in the analysis, but this version is not substantially different from the polynomial-interpolation model. The other variation is a simpler model, a one-variable polynomial fit giving the height of the rocket as a function of time, suitable for a first-semester calculus course.

The remainder of this Module is organized as follows.

- **Section 2**: We provide a list of supplies needed to carry out this experiment and a description of the launch procedure.

- **Section 3**: We present the nine video stills used to generate the raw data for our model and introduce the ground and elevation diagrams needed to convert the apparent position of the rocket as observed from a fixed background.

- **Section 4**: We introduce the trigonometric formulas needed to convert the apparent position of the rocket into the three-dimensional coordinates $X(t)$, $Y(t)$, and $Z(t)$ as a function of time (as measured by the video camera). To generate smooth functions of time for the rocket coordinates, we use Maple 8 to compute the three components of the velocity and acceleration of the rocket (this Module could easily be adapted to any computer algebra system or graphing calculator with curve-fitting capabilities). From these components, we calculate the curvature of the rocket path as a function of time, which exhibits the expected peak near the turn-around point (when the rocket has reached its maximum height). Near the end of the rocket flight, we note an unusual feature in the graph of the curvature, which is explained by a strong gust of wind affecting the path of the rocket near the end of the flight.

- **Section 5**: We briefly discuss modifications of the curve-fitting model used in **Section 4**.

- **Section 6**: We comment on the validity of the zero-torsion model itself and suggest possible augmentation of this experiment.

- **Resources and References**: Additional video stills and Maple code are available from the authors, and we provide annotated references for curve-fitting techniques and for exploring the mathematics of model rockets in general.

2.   Supplies and Procedure

2.1   Supplies Needed

- water rockets (obtain extra in case of defective rockets or cracking on landing); buy "water-powered rockets" from a local toy or hobby store, or order them on the Internet (see **Resources and References**)

- a metric tape measure or a laser telemetry device (if available)

- blueprints for a nearby three-story building (if available—at least the basic dimensions of the building must be known)

- camcorder with videotape

- editing software to view the video frame by frame if available, or at least VCR with a pause button

- stopwatch if unable to view the video frame by frame

2.2   Launching Procedure

• Choose an appropriate backdrop, such as a building with known height. This project is more interesting if the building is only about three stories tall. The rocket will then appear to rise above the building, so that the maximum height has to be estimated by using calculus.

• Before launching the rocket, measure ground distances from the camcorder site to the launch site and from the launch site to the building. The camcorder and launch site should be in line with an easily identifiable location on the building. Measure the distance from the center of the camera lens to ground level. Check if "ground level" at the base of the camera and "ground level" at the base of the building are the same and adjust if necessary.

• Launch the rocket and record the flight on the camcorder. Preview a flight to be sure that the rocket is visible against the building in the video stills. A successful launch is one in which the rocket appears in front of the building both at the beginning and at the end of the flight and appears to rise above the roofline at its maximum height. Several launches may be necessary to achieve this result. Having someone say, for example, "This is the third trial" as you begin to record the launch will help identify the different trials when viewing the tape later.

• Once the rocket has landed, measure the ground distances from the camcorder to the landing site and from the launch site to the landing site.

• View the videotape of the successful launches, decide which one to model, and then gather at least nine data points on the path of the rocket, including the launching and landing points. Although an approximating curve could be determined from fewer points, more data gives greater flexibility in choosing which points to generate the curve in the case of the polynomial interpolation fit and a more accurate curve in the case of the least-squares fit. The blueprints of the building in the background will help determine the position of the rocket with respect to the building. If blueprints are unavailable, then make estimates based on the known height of the building and take measurements for the horizontal distances. If you are able to view the tape frame by frame, the fact that most camcorders record at about 30 frames per second can be used to determine timing. Otherwise, use a stopwatch and the pause button to estimate as well as possible.

2.3   Notes on Measurements

• We measure the distances in meters (m), time in seconds (s), and angles in radians.

• Measurement error can be significant in this experiment. If possible, take each measurement twice and average. With three significant digits (the likely limitation of measurements with the available equipment), each measurement has an implicit measurement uncertainty between 0.1% and 1%.

- The standard color video rate is 29.97 frames/sec, which we take to three significant digits as 30.0 frames per second [Compesi 2003].

2.4 Recording Procedure

Once the experiment is completed and a launch chosen, view the videotape frame by frame and extract information about the apparent position of the water rocket during the entire flight path. Construct **Table 1** with careful attention to proper labeling, significant digits, and units. Determine the entries for **Table 2** and present them in a similar form. Develop a model for the rocket flight using either polynomial interpolation, least-squares fit, or the one-dimensional model. Analyze the model, being sure to address all the questions raised in **Section 4.3**.

3.   The Water Rocket Flight

3.1 Building Blueprint and Video Stills

Figure 2 shows the nine video stills that were the raw data for the model. The rocket was launched from ground level, and the camcorder was 1.52 m above ground level.

Figure 3 shows the exterior blueprints of the library on our campus that we used as a backdrop for our experiment. The scale of the blueprints let us estimate the rocket's horizontal and vertical position at various points during its flight. The initial and terminal positions of the rocket, as well as seven in-flight data points extracted from the video stills, are indicated in the figure.

3.2 Apparent Position of Data Points

The apparent locations of the rocket against the building from the point of view of the camcorder operator are listed in **Table 1**. The heights V_i are measured up from ground level, and the horizontal distances H_i are measured to the right of the point at the intersection of a perpendicular from the camcorder to the building (this point is marked P in **Figure 4**). Also, the times t_i are determined using a video rate of 30.0 frames/sec.

3.3 Ground and Elevation Diagrams

Figure 4 is an aerial view of the rocket launch, showing the relative positions of the camcorder, building, and rocket launch and landing sites.

0th (launch) frame or $t_1 = 0.000$ s.

7th frame or $t_2 = 0.233$ s.

11th frame or $t_3 = 0.367$ s.

15th frame or $t_4 = 0.500$ s.

50th frame or $t_5 = 1.67$ s.

55th launch frame or $t_6 = 1.83$ s.

59th launch frame or $t_1 = 1.920$ s.

63rd launch frame or $t_1 = 2.10$ s.

71st launch frame or $t_1 = 2.37$ s.

Figure 2. Video stills from a successful launch.

Table 1.

Observations.

Frame	i	t_i (s)	H_i (m)	V_i (m)
0	1	0.000	0.000	0.000
7	2	0.233	0.610	5.49
11	3	0.367	1.22	9.45
15	4	0.500	1.83	12.8
50	5	1.67	8.53	13.1
55	6	1.83	9.14	10.1
59	7	1.97	9.75	7.32
63	8	2.10	10.4	4.57
71	9	2.37	16.7	0.000

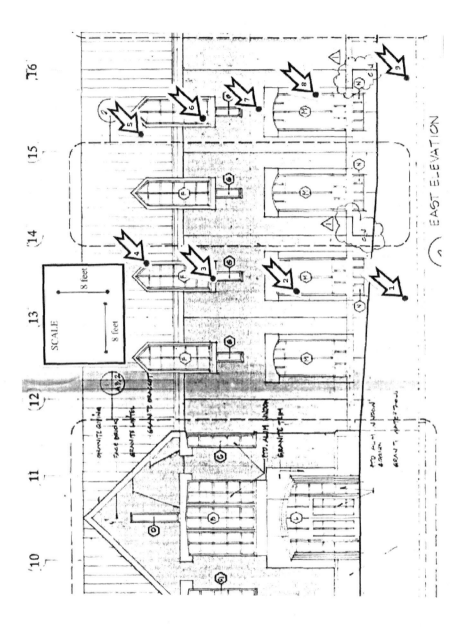

Figure 3. Exterior blueprints of the library used as backdrop for the launch.

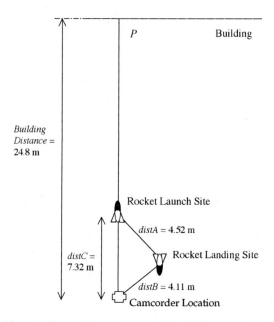

Figure 4. Ground diagram for the rocket flight, *not* to scale.

The data in **Table 1** give only the apparent position of the rocket against the building. We use the similar triangles in **Figure 5** to find the actual heights z_i of the rocket from the perceived heights V_i of the rocket against the building.

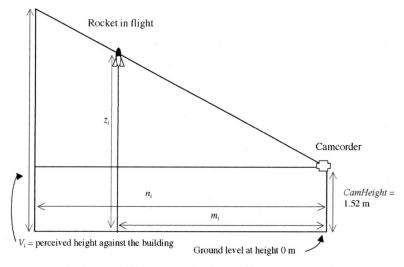

Figure 5. Elevation diagram for the rocket flight, *not* to scale.

4. Developing and Analyzing the Model

4.1 Estimating Rocket Coordinates

We now use basic trigonometry to estimate the three-dimensional coordinates (x_i, y_i, z_i) of the rocket from the perceived positions (H_i, V_i) of the rocket relative to the building. **Figure 6** illustrates the quantities that we must determine to approximate the x_i and y_i ground coordinates of the rocket at time t_i. Note again that our zero-torsion model places the depth coordinate y on the straight line joining the launch site and the landing site.

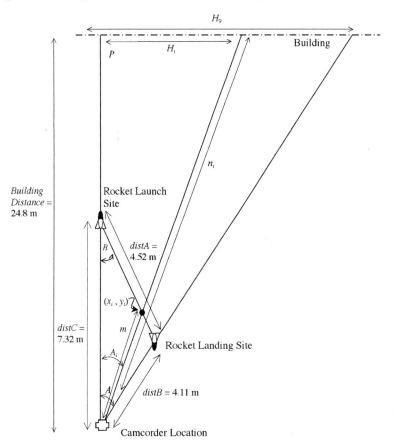

Figure 6. Location diagram for the rocket flight, *not* to scale.

Table 2 gives the quantities used and the coordinates calculated. We find the times t_i by dividing the frame number by 30.0 frames/sec. At the moment of launch ($t = 0$), the horizontal and vertical distances are 0. Next, use the law

Table 2.

Intermediate values and estimated coordinates.

i	Intermediate Values			Estimated Coordinates			
	A_i (radians)	n_i (m)	m_i (m)	t_i (s)	x_i (m)	y_i (m)	z_i (m)
1	0.0000	24.8	7.32	0.000	0.000	7.32	0.00
2	0.0245	24.8	7.02	0.233	0.172	7.02	2.64
3	0.0490	24.9	6.76	0.367	0.331	6.75	3.68
4	0.0735	24.9	6.52	0.500	0.479	6.50	4.48
5	0.331	26.3	4.88	1.67	1.59	4.62	3.68
6	0.353	26.5	4.79	1.83	1.66	4.50	3.07
7	0.374	26.7	4.71	1.97	1.72	4.39	2.55
8	0.395	26.9	4.64	2.10	1.79	4.28	2.05
9	0.591	29.9	4.11	2.37	2.29	3.42	0.00

of cosines to find angle A, the angle at the camcorder between the rocket launch and landing sites:

$$A = \arccos\left(\frac{(distB)^2 + (distC)^2 - (distA)^2}{2 \times (distB)(distC)}\right) = 0.591 \text{ radians} = 33.9°.$$

Since the line from the camcorder through the launch site forms a right angle with the building, this angle can be used to find the horizontal distance H_9 along the building at landing:

$$H_9 = (BuildingDistance)\tan A = 16.7 \text{ m}.$$

Use the horizontal distances H_i from **Table 1** to find the ground angles between the launch site and the rocket position at each time (see **Figure 6**). To find the other angles listed in **Table 2**, use

$$A_i = \arctan\left(\frac{H_i}{BuildingDistance}\right) \text{ radians}.$$

Also, use A and the fact that the line from the camcorder through the launch site to the building is perpendicular to the building to find the hypotenuses n_i (see **Figure 6**):

$$n_i = \frac{BuildingDistance}{\cos A_i}.$$

Next, use the law of cosines to find B, the angle at the launch site from the camcorder to the landing site:

$$B = \arccos\left(\frac{(distB)^2 + (distC)^2 - (distA)^2}{2 \times (distB)(distC)}\right) = 0.532 \text{ radians} = 30.5°.$$

Use the proportions from the law of sines to find the lengths of m_i (see **Figure 6**):

$$m_i = \frac{(distC)\sin B}{\sin\left(\pi - (B + A_i)\right)}.$$

Assuming that the rocket's path remains in the vertical plane containing its launch and landing sites (i.e., the zero-torsion model), the A_i and m_i values give the angles and magnitudes for the projection into the X-Y plane of the position vectors for the rocket at time t_i:

$$x_i = m_i \cos\left(\tfrac{\pi}{2} - A_i\right), \qquad y_i = m_i \sin\left(\tfrac{\pi}{2} - A_i\right).$$

Now, use similar triangles to approximate the actual rocket heights from the perceived heights against the building (see **Figure 5**). First, define $z_1 = 0$ and $z_9 = 0$ as the launch and landing heights. Next, estimate the remaining actual heights using similar triangles. Remember to subtract the camcorder height from the V_i and then add it back in to get the actual rocket heights z_i. Thus, we get:

$$z_i = CamHeight + \frac{(V_i - CamHeight)(m_i)}{n_i}.$$

We plot the resulting points (x_i, y_i, z_i) in three-dimensional space in **Figure 7**.

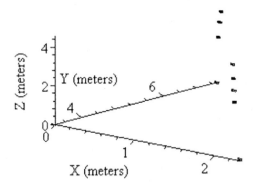

Figure 7. Points on the path of the rocket, from launch on the Y-axis to landing on the X-axis, *not to scale.*

4.2 Modeling the Flight Path Using Polynomial Interpolation

With estimates of the three-dimensional coordinates for the rocket at nine different times during its flight, we can create a space curve modeling the flight path. We use seven of the nine data points (omitting the in-flight points at times t_3 and t_6) to create a sixth-degree interpolating polynomial in the X, Y, and Z

components. We use seven points, because to determine all of the coefficients (including the constant term) uniquely, the number of points must be exactly one more than the degree of the polynomial. We omitted the third and sixth data points after experimenting with omitting different pairs of points to see which yielded the best model. In the absence of more sophisticated curve fitting techniques, this experimentation is an important part of the modeling process.

The resulting sixth-degree interpolating polynomials obtained from our zero-torsion model for the X, Y, and Z coordinates (expressed in meters, with time in seconds) are:

$$X(t) = \quad\quad 0.0412t + 4.65t^2 - 8.73t^3 + 7.63t^4 - 3.12t^5 + 0.480t^6,$$
$$Y(t) = 7.32 - 0.0701t - 7.90t^2 + 14.8\ t^3 - 13.0t^4 + 5.31t^5 - 0.816t^6,$$
$$Z(t) = \quad\quad 15.0t - 20.8t^2 + 27.3\ t^3 - 24.3t^4 + 10.3t^5 - 1.59\ t^6.$$

We graph them separately in **Figure 8** and together as a parametrized curve in three-space in **Figure 9**.

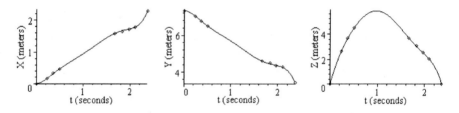

Figure 8. Graphs of X, Y, and Z.

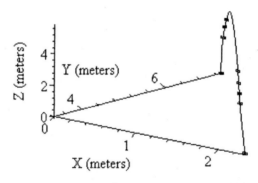

Figure 9. Three-dimensional parametrized curve of the flight path.

As in each separate coordinate, the interpolating curve fits all of the data points well, even the points corresponding to times t_3 and t_6. The coefficients of the sixth-degree interpolating polynomials are not required to possess physical interpretations, in contrast to, for example, a least-squares fit of the flight path based on a parabolic model.

The polynomial fit curve shows the effect of the gusting wind on the empty (and hence light) rocket toward the end of its flight. It also gives a means of estimating the maximum height attained by the rocket, a quantity which could not be determined from the raw data alone. This and other applications of the model are addressed in next section.

4.3   Analysis of the Flight

The Requirements of this ILAP are to answering some of the many questions that naturally arise about the rocket flight path:

1. How high did the rocket go?

2. How fast was it going at different times? How much of that speed was due to vertical motion, and how much to lateral motion? What was its acceleration at different times?

3. How far did the rocket travel during its flight?

4. How sharply did it turn around when it reached its apex? Was that the tightest turn it made during the flight?

5. When did the three stages (thrusting, coasting, recovery) of the flight occur?

6. To what extent did the wind counteract gravity?

7. How much did the wind blow the rocket off the parabolic path predicted by the standard model?

From just the video stills and the resulting data, we cannot address any of these questions. However, now that we have a model approximating the rocket flight, we can apply calculus concepts to find reasonable answers for these questions.

The analysis below illustrates the ways that our curve may be physically interpreted. Another rocket flight will undoubtedly lead to a different-shaped curve with different physical properties. For example, in response to question 6, the graph shows that a wind gust with considerable horizontal strength actually lifted the rocket toward the end of its flight. A significantly different curve may result from other wind effects, or a windless day, and consequently must be considered in its own right rather than just mimicking our analysis.

To address these questions, we use the space curve $\mathbf{R}(t)$ modeling the rocket flight that we found using sixth-degree polynomial interpolation:

$$\mathbf{R}(t) = \langle X(t), Y(t), Z(t) \rangle.$$

We also need the velocity vector $\mathbf{R}prime(t)$ (with units of m/s) and acceleration vector $\mathbf{R}2prime(t)$ (with units of m/s^2), vector-valued curves that can be found by taking the component-wise first and second directives of $R(t)$, respectively:

$$\mathbf{R}prime(t) = \langle X'(t), Y'(t), Z'(t) \rangle, \qquad \mathbf{R}2prime(t) = \langle X''(t), Y''(t), Z''(t) \rangle.$$

Specifically, we have

$$X'(t) = \quad 0.0412 + 9.29t - 26.2t^2 + 30.5t^3 - 15.6t^4 + 2.88t^5,$$
$$Y'(t) = \quad -0.0701 - 15.8\ t + 44.5t^2 - 51.9t^3 + 26.6t^4 - 4.90t^5,$$
$$Z'(t) = \quad 15.0 \quad -41.6\ t + 81.8t^2 - 97.4t^3 + 51.4t^4 - 9.57t^5,$$

and

$$X''(t) = \quad 9.29 - \quad 52.4t + \quad 91.6t^2 - \quad 62.5t^3 + 14.4t^4,$$
$$Y''(t) = -15.8 \quad + \quad 89.0t - 156\ \ t^2 + 106\ \ t^3 - 24.5t^4,$$
$$Z''(t) = -41.6 \quad + 164\ \ t - 292\ \ t^2 + 206\ \ t^3 - 47.8t^4.$$

We use the velocity and acceleration vectors also to calculate the Frenet-Serret curvature for the three-dimensional parametrized space curve.

Requirement 1: How high did the rocket go?

This is perhaps the most natural question to ask that cannot be readily answered simply by watching the videotape. However, now that we have a smooth flight-path curve, finding the maximum height of the rocket is easily done by setting the vertical component of the velocity equal to zero and solving for $t = t_{max}$, the time when the rocket reached its apex. We then find the maximum height by substituting t_{max} into $Z(t)$:

$$Z'(t) = 15.0 - 41.6t + 81.8t^2 - 97.4t^3 + 51.4t^4 - 9.57t^5;$$
$$Z'(t) = 0 \text{ when } t = 0.974 \text{ s and } Z(0.974) = 5.81 \text{ m}.$$

A maximum height of about 5.81 m is not bad for a water-propelled rocket about 13 cm long! However, in the absence of air resistance and with an initial vertical speed of 15.0 m/s, the rocket should have reached a maximum height of $(15.0)^2/(19.6) = 11.5$ m, or almost twice as high as our model predicts. One can readily see the effect of air resistance on the maximum height.

Requirement 2: How fast was the rocket going at the launch, apex, and landing times? How much of that speed was due to vertical motion, and how much to lateral motion? What was its acceleration at those times?

To address these questions, we find and analyze the velocity and acceleration vectors and their magnitudes. We begin first with the components of the velocity vectors, with each component measured in m/s:

$$\mathbf{R}'(0) = \langle 0.0412, \ -0.0701, \quad 15.0 \rangle,$$
$$\mathbf{R}'(t_{max}) = \langle \ 0.937, \quad -1.59, \quad 0 \rangle,$$
$$\mathbf{R}'(t_{land}) = \langle \ 3.88, \quad -6.59, \ -13.5 \rangle,$$

where $t_{land} = t_9 = 2.37$ s. The signs of the components of each velocity vector indicate direction, information that is lost when we compute the overall speed below. At the launch, the derivatives in the X and Y components are very small (in absolute value), reflecting the fact that we actually did a pretty good job of launching the rocket vertically (i.e., the initial launch angle is $\theta = 89.7°$ as measured from the horizontal), and that the later lateral motion must have been due to the wind. At the apex of the flight path, even though the $Z(t)$ derivative is zero, the $X(t)$ and $Y(t)$ derivatives indicate that the rocket was still moving laterally. Upon impact, although the rocket hit the ground with nearly the same vertical velocity as it left the ground, it was still moving quite fast laterally.

To see the overall speed at these three times, we need the magnitudes of the velocity vectors. We find that their magnitudes are

$$\text{speed}(0) \quad = |\mathbf{R}'(0)| \quad = 15.0 \ \ \text{m/s} = 33.5 \ \ \text{mph},$$
$$\text{speed}(t_{\max}) = |\mathbf{R}'(t_{\max})| = \ \ 1.85 \ \text{m/s} = \ \ 4.14 \ \text{mph},$$
$$\text{speed}(t_{\text{land}}) = |\mathbf{R}'(t_{\text{land}})| = 15.5 \ \ \text{m/s} = 34.6 \ \ \text{mph},$$

where we use the conversion 1 mph = 0.447 m/sec.

Comparing these speeds to the values of the individual components shows that the speed at launch was almost entirely in the vertical direction, whereas when the rocket landed, because of the lateral motion, it was moving faster.

Now, consider the acceleration vectors at these three times, with each component measured in m/s^2:

$$\mathbf{R}''(0) = \langle \ 9.29, \quad -15.8, \ -41.6 \rangle,$$
$$\mathbf{R}''(t_{\max}) = \langle 0.418, \ -0.711, \ -12.4 \rangle,$$
$$\mathbf{R}''(t_{\text{land}}) = \langle \ 21.8, \quad -37.1, \ -65.3 \rangle.$$

The negative value of the acceleration in each Z component illustrates the general property that when an object is slowing down, the acceleration is in the opposite direction of the motion, and when it is speeding up, the acceleration is in the same direction as the motion. Thus, when rising and slowing down, the rocket has negative acceleration (in opposite direction to its ascent), and when falling and speeding up, the rocket also has negative acceleration (in the same direction as its descent).

Next, we find the magnitude of the acceleration vector at these three times, each again measured in m/s^2. We can then compare these values to some common known accelerations.

$$\text{acceleration}(0) = |\mathbf{R}''(0)| \quad = 45.4,$$
$$\text{acceleration}(t_{\max}) = |\mathbf{R}''(t_{\max})| = 12.4,$$
$$\text{acceleration}(t_{\text{land}}) = |\mathbf{R}''(t_{\text{land}})| = 78.2.$$

The starting acceleration is about 50% more than that of a Space Shuttle at take-off (29 m/s^2), the apex acceleration is about 50% more than that of a cheetah at takeoff (7.8 m/s^2), and the landing acceleration is about that of a parachute at landing (35 m/s^2) [Vawter 2003]. We quickly point out, however, that although the acceleration at apex is relatively close to the theoretical prediction of 9.8 m/s^2 for g, the landing acceleration indicates that the rocket hit the ground with an acceleration of about $8g$, which is physically impossible in the absence of wind effects.

Considering the effects of air resistance and gravity alone, we conclude on physical grounds that the magnitude of the vertical component of the acceleration should be larger than the gravitational acceleration ($g = 9.81$ m/s^2) during ascent (when gravity and air resistance are in the same direction), while it should be less than the gravitational acceleration during descent (when gravity and air resistance are in opposite directions). Although the numerical results for the acceleration show a vertical component relatively close to g at the apex, the values at launch and landing times raise serious doubts about the numerical validity of the model.

Requirement 3. How far did the rocket travel during its flight? Looking at the flight path, it seems that the rocket went farther coming down than going up, but how much farther?

We can answer this question by finding and interpreting the arclength of the flight path, over various time intervals. First, we determine the length over the entire flight:

$$\int_0^{t_{\text{land}}} |\mathbf{R}'(t)|\, dt = \int_0^{t_{\text{land}}} \sqrt{[X'(t)]^2 + [Y'(t)]^2 + [Z'(t)]^2}\, dt = 12.8 \text{ m}.$$

Then, we determine the lengths over the rising and falling parts of the flight:

$$\int_0^{t_{\text{max}}} |\mathbf{R}'(t)|\, dt = \int_0^{t_{\text{max}}} \sqrt{[X'(t)]^2 + [Y'(t)]^2 + [Z'(t)]^2}\, dt = 6.26 \text{ m},$$

$$\int_{t_{\text{max}}}^{t_{\text{land}}} |\mathbf{R}'(t)|\, dt = \int_{t_{\text{max}}}^{t_{\text{land}}} \sqrt{[X'(t)]^2 + [Y'(t)]^2 + [Z'(t)]^2}\, dt = 6.54 \text{ m}.$$

The rocket traveled about 12.8 m, about 6.26 m during the first part of the flight and 6.54 m during the second part (when the rocket was lighter and the wind blew it more off course). In fact, taking into account the effects of air resistance and gravity alone, we would expect on physical grounds that the distance covered during descent should be shorter than the distance covered during ascent. Hence, the longer distance during the descent is indeed an indication of a wind gust with significant in-plane strength.

Requirement 4. How sharply did the rocket turn around when it reached its apex? Is this the tightest turn it made during its flight?

The Frenet-Serret curvature function, $\kappa(t)$, helps answer this question. Recall that the radius of the osculating circle is the reciprocal of κ; so the larger the value of κ is, the tighter the curve is:

$$\text{curvature} = \kappa(t) = \frac{|\mathbf{R}'(t) \times \mathbf{R}''(t)|}{|\mathbf{R}'(t)|^3}.$$

The curvature at the apex, when $t = t_{max} = 0.974$ s, is $\kappa(t_{max}) = 3.62$ m^{-1}. Thus, the osculating circle has radius $1/\kappa(t_{max}) = 0.276$ m, which is about the size of an extra large pizza.

But is this necessarily the maximum curvature? Consider the graph of the curvature (**Figure 10**).

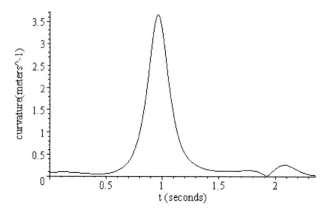

Figure 10. Graph of curvature vs. time.

Since $\kappa'(t) = 0$ when $t = 0.963$ s, the maximum curvature is $\kappa(0.963) = 3.65$ m^{-1}, with osculating circle radius of 0.274 m. Therefore, it looks as if the maximum curvature occurred slightly after the rocket turned around at its apex. Since the values are so close, it is difficult to determine if this is due to the wind pushing sideways as the rocket slowed, and thus "loosening" the curve, or whether it is due to the vagaries of the model.

We notice a remarkable feature of curvature versus time near the 1.9-s mark. There the curvature becomes very small, indicating that the velocity and acceleration become nearly parallel for a short time. This is possible if a wind gust with significant planar strength appeared at that time.

Requirement 5. When did the three stages of thrusting, coasting, and recovery of the rocket flight occur?

While it would be quite difficult to determine the answer by watching the videotape (especially the transition from thrusting to coasting), and almost impossible to determine it from the nine data points, the stages can be quite clearly seen from the plots of the magnitudes of the velocity and acceleration functions.

First, we plot in **Figure 11** the three components of the velocity curve.

Then we find the speed (magnitude of the velocity) and plot the resulting curve (**Figure 12**).

$$\text{speed}(t) = |\mathbf{R}'(t)| = \sqrt{[X'(t)]^2 + [Y'(t)]^2 + [Z'(t)]^2}.$$

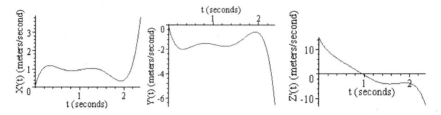

Figure 11. The three components of the velocity curve.

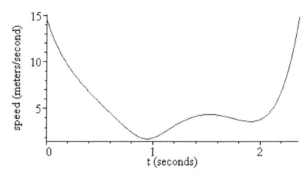

Figure 12. Speed vs. time.

The speed curve gives a preliminary indication of the stages. We already know the transition point between the coasting and recovery stages, since that is just the time when the rocket reached it maximum height at $t_{max} = 0.974$ s.

The more subtle question is the transition between the thrusting and coasting stages. Note that the rocket starts out going quite fast, and then the speed decreases. There is a slight bend in the speed curve around 0.3 seconds, but it is hard to see exactly what is happening when. The magnitude of the acceleration will give us more information. First, we plot the three components of the acceleration curve (**Figure 13**).

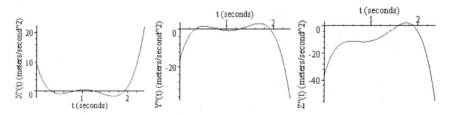

Figure 13. The three components of the acceleration curve.

Then we find and plot the acceleration magnitude curve (**Figure 14**).

$$|\text{acceleration}(t)| = |\mathbf{R}''(t)| = \sqrt{[X''(t)]^2 + [Y''(t)]^2 + [Z''(t)]^2}.$$

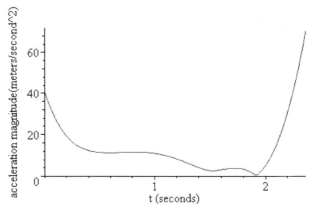

Figure 14. Acceleration magnitude vs. time.

We can see the transition from the thrusting stage to the coasting stage more clearly in the acceleration magnitude graph. The rocket was accelerating (although less and less as it lost water and pressure) from launching to somewhere around 0.5 s and then leveled off somewhat during the coasting stage. More specifically, we have

$$\frac{d}{dt}|\text{acceleration}(t)| = \frac{d}{dt}|\mathbf{R}''(t)| = 0 \text{ when } t = 0.518 \text{ s.}$$

Between 1.8 s and 2 s, apparently a strong gust of wind caught the rocket and forced it sideways and down. The end of the coasting stage is when the rocket turns around at t_{max}, and the rest of the flight is the recovery.

Therefore, the thrusting went from $t = 0$ s to $t = 0.518$ s, the coasting from $t = 0.518$ s to t_{max}, and the recovery from t_{max} to t_{land}.

Requirement 6. To what extent did the wind counteract gravity?

Since gravity is in the vertical direction, it suffices to consider just the vertical component of the acceleration vector, and we need to compare this to the constant acceleration due to gravity (-9.81 m/s²), which otherwise would be the only force acting on the rocket after the thrusting stage finished.

In the absence of wind effects or air resistance, we would have expected $Z''(t)$ to increase to -9.81 m/s² during the thrusting stage and then remain constant for the rest of the flight. Any deviation from this thus must reflect the effects of wind gusts and/or air resistance. **Figure 15** indicates that the upward force from the wind actually exceeded the downward force of gravity.

Setting $Z''(t) = 0$, we find that $t = 1.57$ s to $t = 1.90$ s was the interval during which this occurred. Also, toward the end of the flight, a gust of wind actually pushed the rocket fairly hard into the ground. However, there was comparatively little wind effect during the coasting stage, as evidenced by the curve staying fairly close to the gravitational constant. Also, compare the

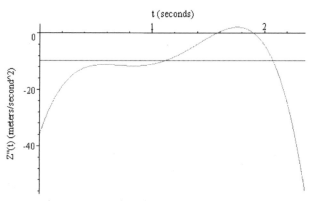

Figure 15. Vertical acceleration vs. time.

two times when $Z''(t)$ is zero to approximately the same times of the two "dips", the second one quite sharp, in the previous acceleration magnitude graph (**Figure 14**).

Requirement 7. How much was the rocket blown off course by the wind?

One way of getting a sense of this is to compare our model to the parameterized curve $\langle (S\cos\theta)t, (S\sin\theta)t - \frac{1}{2}(9.81)t^2 \rangle$, which gives the trajectory of a projectile with no other forces acting on it except gravity. Here S is the initial speed, θ is the launch angle, g is the gravitational constant (-9.81 m/s^2), and the initial position at $t = 0$ is $(x, z) = (0, 0)$. Of course, this does not model our thrusting stage, but it is the best general model that we have.

Since the standard model is in two dimensions, we first need to reparametrize our model along the ground path from launch to landing to get a two-dimensional curve. Thus, the first coordinate function is $\sqrt{[X(t) - X(0)]^2 + [Y(t) - Y(0)]^2}$ and the second coordinate function is simply $Z(t)$.

To use the standard model, we need values for the initial speed S and launch angle θ. We start by estimating them from the t_1 and t_2 data points. Then S is just the change in distance divided by the change in time, and θ is the launch angle from the launch point at $t = t_1 = 0$ to the position at $t = t_2$:

$$S = \frac{\sqrt{[X(t_2) - X(t_1)]^2 + [Y(t_2) - Y(t_1)]^2 + [Z(t_2) - Z(t_1)]^2}}{t_2 - t_1} = 11.4 \,\text{m/s},$$

$$\theta = \arccos\left(\frac{\sqrt{[X(t_2) - X(t_1)]^2 + [Y(t_2) - Y(t_1)]^2}}{\sqrt{[X(t_2) - X(t_1)]^2 + [Y(t_2) - Y(t_1)]^2 + [Z(t_2) - Z(t_1)]^2}} \right)$$

$$= 1.44 \,\text{radians} = 82.5°.$$

The initial velocity is less than the "instantaneous" initial speed of 15.0 m/s found in **Requirement 2** and reflects the changing acceleration during the thrusting stage.

We find the appropriate t ranges by assuming that the second coordinate remains positive. Thus, we solve:

$$(S\sin\theta)t - \tfrac{1}{2}(9.81)t^2 = 0.$$

The two solutions are

$t =$ standard launching time $= 0.000\,\text{s}$, $t =$ standard landing time $= 2.31\,\text{s}$.

The difference in the landing positions gives us a sense of how far "off course" the rocket was blown:

landing difference $= DistA - (S\cos\theta) \times$ (standard landing time)

$= 4.52\,\text{m} - 3.37\,\text{m} = 1.15\,\text{m}.$

Without wind, the standard model estimates that the landing site would have been about 3.37 m away from the launching site, but the rocket actually landed 4.52 m away, a difference of 1.15 m.

We plot and compare our model to the standard parabolic model (**Figure 16**).

It might be argued that the launch angle was actually directly up and so the vertical deviation was due to wind. Furthermore, the rocket was accelerating from t_1 to t_2, so the S and θ values used above may not be the best parameter values for a good comparison. If the initial value shifts at $t = t_i$ to (x_i, z_i), the model becomes

$$\langle (S\cos\theta)(t - t_i) + x_i, \quad (S\sin\theta)(t - t_i) - \tfrac{1}{2}g(t - t_i)^2 + z_i \rangle.$$

We estimated the parameters S and θ using times t_2 and t_3, since the thrust was diminished by then. **Figure 17** gives the plot of our model vs. the adjusted standard model. The figure illustrates that there was a thrusting phase occurring during the actual rocket flight. Also, while this two-dimensional model involves no wind effect, it gives a greater ground distance than was actually covered by the rocket. Since the rocket was in fact blown by the wind, this is probably a less useful model than the parabolic model derived using t_1 to t_2.

5.   Model Variations

5.1   Modeling the Flight Path Using Least-Squares

We can also find a viable model through a least-squares fit of all nine data points of **Table 1**. The model created using a sixth-degree least-squares fit is quite similar in shape to the sixth-degree polynomial interpolation of **Section 4.2**. For example, both models achieve similar maximum height. Preference for one method over the other may depend upon which modeling tools are available, where this Module is integrated in the curriculum, and mathematical background.

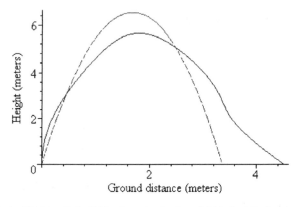

Figure 16. Actual flight path (solid line) vs. standard model (dashed line), using t_1 and t_2.

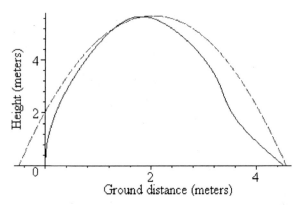

Figure 17. Actual flight path (solid line) vs. adjusted standard model (dashed line), using t_2 and t_3.

Using the times t_i and coordinates x_i, y_i, and z_i from **Table 2**, we can find a least-squares fit for the X, Y, and Z components. The resulting sixth-degree least-squares polynomials for the X, Y, and Z components, measured in meters, are:

$$X(t) = \qquad\ 0.161t + \ 3.80t^2 - \ 6.74t^3 + \ 5.77t^4 - 2.37t^5 + 0.369t^6,$$
$$Y(t) = 7.32 - \ 0.277t - \ 6.46t^2 + 11.5\ t^3 - \ 9.81t^4 + 4.03t^5 - 0.627t^6,$$
$$Z(t) = \qquad 14.6\quad t - 18.1\ t^2 + 20.7\ t^3 - 18.1\ t^4 + 7.74t^5 - 1.22\ t^6.$$

We can then combine the least-squares fits in the three components to create a space curve to model the rocket's flight (**Figure 18**).

As with the interpolated version, this curve fits the data points well and can be used to analyze the rocket flight. It gives slightly different values for the questions considered in **Section 4.3**. For example, to estimate the maximum

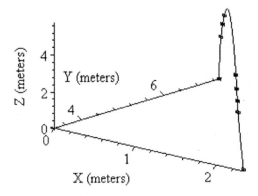

Figure 18. Space curve model of the rocket's flight from sixth-degree least-squares fit to the data.

height, we compute the derivative of $Z(t)$, and find the critical value of $Z(t)$:

$$Z'(t) = 14.6 - 36.2t + 62.0t^2 - 72.4t^3 + 38.7t^4 - 7.32t^5,$$
$$Z'(t) = 0 \text{ when } t = 0.958 \text{ s and } Z(0.958) = 5.68 \text{ m}.$$

Thus, according to our least-squares curve, the rocket reached its maximum height approximately 0.958 s into the flight and the height was approximately 5.68 m, which is less than the 5.81 m approximation found with the interpolated curve in **Section 4.3**.

5.2   Simpler Height vs. Time Model

It is also possible to use calculus of one variable to analyze the rocket flight from the height function alone. Much of the work is very similar to that in the three-dimensional case. Estimating the actual rocket positions is done as in **Section 4.1**, and finding a model $Z(t)$ for the height as a function of time uses interpolation as in **Section 4.2**. The functions $X(t)$ and $Y(t)$ do not need to be determined. **Figure 19** shows the data points and the curve $Z(t)$.

Since the height function $Z(t)$ is identical to the third (vertical) component of the three-dimensional model, some of the analysis below repeats what was presented earlier. We include this as necessary to provide a complete consideration of the height vs. time model.

The height function alone does not model the strong lateral drift due to the wind at the end of the flight, but it does show the slight lift and sharp drop at the end of the flight as the wind forced the rocket down.

As in **Section 4.3**, we can address some questions about the flight, using the first and second derivatives of $Z(t)$:

$$Z'(t) = \quad 15.0 - 41.6t + 81.8t^2 - 97.4t^3 + 51.4t^4 - 9.57t^5,$$
$$Z''(t) = -41.6 + 164t - 292t^2 + 206\,t^3 - 47.8t^4 \qquad .$$

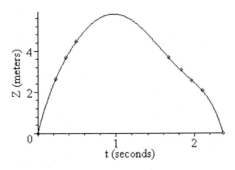

Figure 19. Model of the rocket's height from sixth-degree least-squares fit to the data.

Requirement 1. How high did the rocket go?

Setting $Z'(t) = 0$ and solving the time t_{max} when the rocket reached its apex. We find the maximum height by substituting t_{max} into $Z(t)$:

$$Z'(t) = 0 \text{ when } t = t_{max} = 0.974 \text{ s and } Z(0.974) = 5.81 \text{ m.}$$

Requirement 2. How fast was the rocket going at the launch, apex and landing times? What was its acceleration at those times?

To find the velocities, we substitute the three times into $Z'(t)$:

$$
\begin{aligned}
Z'(0) &= 15.0 \text{ m/s} = 33.5 \text{ mph,} \\
Z'(t_{max}) &= 0.0 \text{ m/s} = 0.0 \text{ mph,} \\
Z'(t_{land}) &= -13.5 \text{ m/s} = -30.1 \text{ mph.}
\end{aligned}
$$

Next, we find the accelerations at these three times, each measured in m/s^2:

$$
\begin{aligned}
Z''(0) &= -41.6, \\
Z''(t_{max}) &= -12.4, \\
Z''(t_{land}) &= -65.3.
\end{aligned}
$$

Requirement 5. When did the three stages, thrusting, coasting and recovery, of the rocket flight occur?

The recovery stage begins at time t_{max}. The end of the thrusting stage (and beginning of the coasting stage) can be more subtle. We first consider the graphs of the velocity $Z'(t)$ and the acceleration $Z''(t)$ functions, plotted together in **Figure 20**.

The thrusting stage ends with the first maximum of $Z''(t)$, so we must determine when this occurs:

$$\frac{d}{dt} Z''(t) = 0 \text{ when } t = 0.517 \text{ s.}$$

Therefore, the thrusting went from $t = 0$ s to $t = 0.517$ s, the coasting from $t = 0.517$ s to $t = t_{max}$, and the recovery from $t = t_{max}$ to $t = t_{land}$. Between

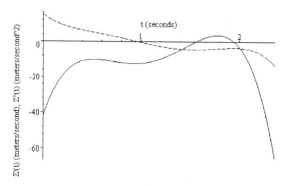

Figure 20. Velocity (dashed line) and acceleration (solid line) vs. time.

1.8 s and 2 s, apparently a strong gust of wind caught the rocket and forced it sideways and down.

Requirement 6. To what extent did the wind counteract gravity?

Since gravity is in the vertical direction, it suffices to consider just the vertical component of the acceleration vector, and we need to compare this to the constant acceleration due to gravity (-9.81 m/s^2), which otherwise would be the only force acting on the rocket after the thrusting stage finished.

If there had been no wind or air resistance, we would have expected $Z''(t)$ to increase to -9.81 m/s^2 during the thrusting stage and then remain constant for the rest of the flight. Any deviation from this thus must reflect these effects. The graph of $Z''(t)$ in **Figure 20** (solid curve) indicates that the upward force from the wind actually exceeded the downward force of gravity. Setting $Z''(t) = 0$, we find that $t = 1.55$ s to $t = 1.95$ s seconds was the interval during which this occurred. Also, toward the end of the flight, a gust of wind actually pushed the rocket fairly hard into the ground. However, there was comparatively little wind effect during the coasting stage as evidenced by the curve staying fairly close to the gravitational constant.

Requirement 7. What effect did the wind have on the height attained by the rocket?

We use the standard parabolic model for the height of a projectile, $s(t) = v_0 t - \frac{1}{2}gt^2$, to get some sense of what effect the wind had on the height.

We need a value for the initial velocity v_0. We start by estimating it from the t_1 and t_2 data points. The velocity v_0 is then approximately the change in distance divided by the change in time:

$$v_0 \approx \frac{Z(t_2) - Z(t_1)}{t_2 - t_1} = 11.3 \text{ m/s}.$$

We find the appropriate t ranges by assuming that standard model remains positive. Thus, we solve $v_0 t - \frac{1}{2}(9.81 t^2) = 0$, getting

$t = 0.000$ s   and $t = $ standard landing time $= 2.31$ s.

We plot and compare our model to the standard parabolic model (**Figure 21**).

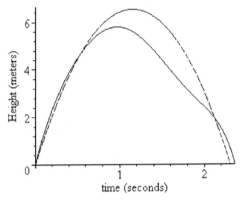

Figure 21. Height model (solid line) from sixth-degree least-squares fit vs. standard parabolic model (dashed line) estimating v_0 from t_1 and t_2.

The greater height of the parabolic model reflects the difference in thrust in the two models.

We look at the difference in the air time, which will give us a sense of how the wind may have held the rocket in the air longer than might be expected, or driven it into the ground earlier than might be expected.

air time difference $= |t_{\text{land}} - $ standard landing time $| = 0.0564$ s.

The rocket was accelerating from t_1 to t_2, so the initial velocity v_0 used above may not be the best parameter for a good comparison. If the initial value shifts to (t_i, z_i), the model becomes $S(t) = v_0(t - t_i) - \frac{1}{2}g(t - t_i)^2 + z_i$. We estimate the parameter v_0 using times t_2 and t_3, since the thrusting stage was about over by then. In this case, we get the plot of our model vs. the standard model as shown in **Figure 22**.

Comparing these two models, we get an air time difference of 0.209 s in this case. This model gives a better fit for the first part of the flight; and since the rocket was in fact blown by the wind, this is probably a more useful comparison than the parabolic model derived using t_1 to t_2.

6. Comments on the Modeling

Hindsight is 20/20. In the process of running this project with our classes, repeating it to get the data presented here, and then writing this Module, we identified problem areas, thought of ideas for improving the project, and discovered new resources. We plan to use some of these items in the future, and offer them here for consideration. We hope that anyone who tries this Module,

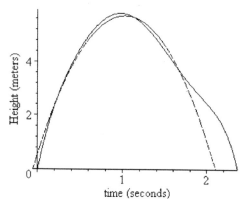

Figure 22. Height model (solid line) from sixth-degree least-squares fit vs. standard parabolic model (dashed line) estimating v_0 from t_2 and t_3.

and especially anyone who experiments with any of the modifications below, will contact us to share their experiences.

6.1 Measurements

The task of taking accurate measurements in this experiment is difficult because of uneven turf and uncertain elevations. These inaccuracies can have a considerable effect on the model. Furthermore, the method used to estimate the launching and landing positions of the rocket (ground measurements) was different from the method used to estimate the in-flight positions (projection against the building). This must be done carefully to avoid misleading results. More sophisticated measuring tools, if available, could help here. For example, the use of laser telemetric devices (e.g., electronic theodolites [Mohave Instrument 2003]) would improve the measurements of distance.

6.2 Sensitivity

It is instructive to see how robust our numerical results are by introducing a measurement uncertainty in our perceived positions of the rocket during its flight path. First, we look at how uncertainties in the position of the water-rocket along its path can change the maximum height achieved by the water rocket. In **Figure 23** (where the parabolic curves are generated from the first two times), we show the influence of a 1% uncertainty (**Figure 23a**) and a 10% uncertainty (**Figure 23b**) artificially imposed on the estimated coordinate data in **Table 2**. A 10% uncertainty, which can result from making measurements to two significant digits, shows tremendous variations in maximum height and, by extension, maximum horizontal distance covered. In contrast, a 1% uncertainty, which can result from making measurements to three significant digits, shows only small variations.

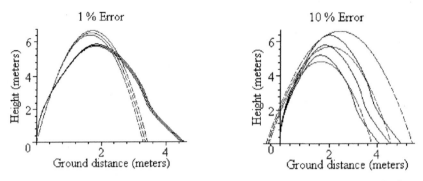

Figure 23. Influence of measurement error on flight path (solid line) and on standard model (dashed line), for 1% error and 10% error.

Next, we look at the effect of measurement uncertainties on curvature, which provided a clear signal for a gust of wind occurring at approximately 1.9 s after the launch. While a 1% uncertainty in the data measurements shows virtually no changes in the curvature graph, a 10% uncertainty (as shown in **Figure 24**) clearly shows variations in the maximum curvature achieved during the flight path of the water rocket. We also note, however, that the wind-gust feature at 1.9 s is fairly robust.

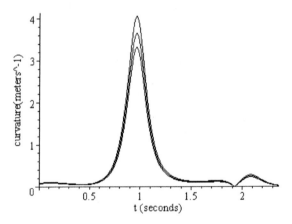

Figure 24. Effect of 10% measurement error on curvature.

6.3   Flight Stages

Because rocket propulsion allows the water rocket to leave the ground with zero initial velocity, the forces acting on the rocket during the thrusting stage are different from those during the coasting and recovery stages. Since the polynomial fit presented here is a *global* fit (i.e., it assumes that the physics of the initial thrusting stage is the same as the later two stages), it mistakenly assigns a nonzero initial velocity. By taking more data points, it should be possible to develop two separate approximating functions, one for the thrusting stage (with zero initial velocity), and one for the coasting and recovery stages (with nonzero initial velocity at the beginning of the coasting stage). This would give a piecewise function as a model of the rocket flight, although presumably a more accurate one. The disadvantage of possibly discontinuous derivatives might be addressed with curve smoothing techniques.

6.4   Weathercocking

Weathercocking is the cumulative effect of the airflow over the nose of the rocket and the wind. This effect can cause the rocket to move into the wind. There is a good description of weathercocking at Folger [2001]. Because the wind was gusting during our launch, we were not able to get a sense of what effect this might have had on our rocket. A water rocket launched during a gentle, steady wind might demonstrate some weathercocking. It would be interesting to try to determine the extent of this effect. It might also be useful to record the wind direction and speed during the flight. The technical difficulties are finding a practical means for doing so and correlating the wind data with the flight data.

6.5   Three-Dimensional Coordinates

We have always done this project using a single video camera. This has forced us to make the simplifying assumption that the rocket remains in the vertical plane through the ground path. Using two widely spaced video cameras to record the launch would allow the construction of a nonplanar space curve. It would require careful synchronization of the recordings. More critically, significantly more work would need to be done to estimate the actual rocket positions from its two different apparent positions against the building from the video stills of the two different cameras. This would shift the bulk of the computation in the project to three-dimensional coordinate geometry and away from the calculus of three variables that we wanted to emphasize. However, it would presumably give a more interesting space curve as a model, and would make considering other characteristics, such as the pitch, yaw, and roll, of the rocket more meaningful. The debate continues, and we would be very glad to hear from anyone who tries this "binocular" version!

6.6   Measuring Maximum Height with a
Simple Sextant

Although the videotape is not much use in estimating the height of the rocket once it has soared above the building, a simple sextant can be used to determine the height at the apex. This device measures the angle of elevation from eye level to the object sighted through the straw. An observer at a known position relative to the rocket launch site measures the angle of elevation to the rocket when it reaches the apex. The projection of the rocket's position onto the ground must be determined, and the distance from there to the observer computed. That distance and the angle of elevation can then be used to estimate the height. We suggest having at least two people make observations and then average their estimates of the angular distance. If the observers shout "Max!" when they make their observations, the sounds of their voices will be recorded on the videotape, which may give a sense of how close to the actual apex the observations were made.

It would be fun to reserve these data initially and construct a model without them. Then they could be used to see how accurately the model predicts the maximum height. If the model does a poor job, these data could be incorporated into the model, and then analysis done on the improved model.

Figure 25 shows a diagram of a simple sextant; the diagram and instructions below (slightly modified) come from Folger [2001].

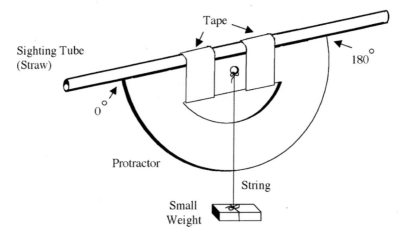

Figure 25. Diagram for a simple sextant.

Materials

- a large-diameter soda straw

- 20 cm length of string

- a protractor

- a weight (an eraser or large washer)

- tape

Construction

Tape the straw across the top of the protractor as shown in the figure. The straw will act as a sighting tube. Secure the string to the protractor by slipping it under the straw and around. Tie the string to itself and tape it to the back of the protractor. Tie the eraser or washer at the opposite end of the string, so that it can act as a weight.

Directions for Use

Look through the straw, focusing on the rocket as it is being launched. Let the weight hang freely but try not to let it swing. Move the device up smoothly as the rocket ascends. At the instant the rocket turns around, hold the string with your finger exactly where it is on the protractor. Record this angle.

Resources and References

Additional Video Stills

We strongly recommend launching and videotaping your own rocket. It is a lot of fun, and much more satisfying to analyze your own data. However, in case that is simply not feasible, we can provide some additional video stills from the same launch as the example in this paper. Please email us for these. There is a scale at the top of the blueprint in **Figure 3** in **Section 3.1** for estimating the apparent position of the rocket against the building.

Maple 8 Code

Also available from us via email request is the Maple 8 code that we used. Here we briefly outline our approach to the code. We first provide generic Maple procedures, which may be used with any rocket flight data set, followed by commands that generate the specific results presented in this Module.

References

Anton, Howard. 1994. *Elementary Linear Algebra*. 7th ed. New York: John Wiley and Sons.

Details on creating interpolated- and least-squares-fit models can be found in Sections 6.4 and 9.3

Benson, Tom. 2002. NASA Glenn Research Center: Beginner's Guide to Aeronautics. http://www.grc.nasa.gov/WWW/K-12/airplane/index.html .

This Website also offers a "Beginner's Guide to Aerodynamics" and a "Beginner's Guide to Model Rockets." The model rocket information includes some of the basic mathematics and physics concepts that affect model rocket design and flight. The site is intended to offer K-12 teachers, students, and others with basic background information on aerodynamics and propulsion.

Compesi, Ronald J. 2003. *Videofield Production and Editing.* 6th ed. Boston: Allyn and Bacon.

Estes Industries. 2003. Publications. `http://www.esteseducator.com/cfusion/publications.cfm` .

Estes Industries for over 40 years has been "the world's leader in the manufacturing of safe, reliable, and high quality rocket products" and offers a variety of "new and innovative educational products." This site is a comprehensive source of information about how model rocketry can be integrated into the classroom; it offers links to a number of different publications (in PDF format) on model rocketry, ranging from order forms to teachers' guides to elementary mathematics information on rocket flights.

Ference, Dave. 2003. Dave's Cool Toys. `http://www.davescooltoys.com/` .

"Dave's Cool Toys" is a Website at which water-powered rockets can be purchased inexpensively. The link sequence `Our Toys ⟶ Outside` leads to ordering information about water-powered rockets as well as higher-flying "meteor rockets" (powered by a baking soda and vinegar solution) and even higher-flying "air burst rocket systems" (powered by compressed air).

Finney, G.A. 2000. Analysis of a water-propelled rocket: A problem in honors physics. *American Journal of Physics* 68: 223–227. `http://ojps.aip.org/journals/doc/AJPIAS-ft/vol_68/iss_3/223_1.html` .

This paper looks at the effects of air resistance and thrust on the one-dimensional path of a water-propelled rocket launched vertically.

Folger, Daniel, and Maria Folger. 2001. The Future Astronauts of America Foundation. `http://www.geocities.com/CapeCanaveral/Hangar/5421/index.html` .

The Future Astronauts of America Foundation is "a multi-faceted program designed to encourage young students and adults to become involved with the exciting world of space science and its applications." This site provides information about model rocket components. It also offers useful descriptions of the flight sequence of a model rocket, including its thrusting, coasting, and recovery stages.

Mohave Instrument Co. 2003. `http://www.mohaveinstrument.com/`.

This site contains information about commercially available electronic theodolites.

Nave, C. Rod. 2000. HyperPhysics. `http://hyperphysics.phy-astr.gsu.edu/hbase/hframe.html`.

This Website is "an exploration environment for concepts in physics which employs concept maps and other linking strategies to facilitate smooth navigation." It contains useful information on rocket propulsion.

Prusa, J.M. 2000. Hydrodynamics of a water rocket. *SIAM Review* 42 (4): 719–726. `http://epubs.siam.org/sam-bin/dbq/article/34822`.

This paper details a general model for a water rocket flight and offers advanced analysis of the trajectory. The paper is suitable for students with applied mathematics or engineering interests, at the advanced undergraduate level or beginning graduate level (for example, in a fluid dynamics course). It also offers a good application for an advanced graduate-level course in computational methods.

Vawter, Richard. 2003. Typical values of acceleration. `http://www.ac.wwu.edu/~vawter/PhysicsNet/Topics/Kinematics/AccelerationValues.html`.

About the Authors

George Ashline received a B.S. from St. Lawrence University, an M.S. from the University of Notre Dame, and in 1994 a Ph.D. in value distribution theory, also from the University of Notre Dame. He has taught at St. Michael's College since 1995. He is a participant in Project NExT, a program created for new or recent Ph.D.s in the mathematical sciences who are interested in improving the teaching and learning of undergraduate mathematics.

Alain Brizard received a B.Sc. from Sherbrooke University and a Ph.D. from Princeton University in 1990 in astrophysical sciences. He was a staff scientist at the Lawrence Berkeley National Laboratory until he came to St. Michael's College in July 2000. His research activities focus on applications of Lagrangian and Hamiltonian methods in fusion, space, and astrophysical plasma physics.

Joanna Ellis-Monaghan received a B.A. from Bennington College, an M.S. from the University of Vermont, and a Ph.D. from the University of North Carolina at Chapel Hill in 1995 in algebraic combinatorics. She has taught at Bennington College, at the University of Vermont, and since 1992 at St. Michael's College. She is a proponent of active learning and has developed materials, projects, and activities to augment a variety of courses.

Guide for Authors

Focus

The UMAP Journal focuses on **mathematical modeling and applications of mathematics at the undergraduate level.** The editor also welcomes expository articles for the On Jargon column, reviews of books and other materials, and guest editorials on new ideas in mathematics education or on interaction between mathematics and application fields. Prospective authors are invited to consult the editor or an associate editor.

Understanding

A manuscript is submitted with the understanding—unless the authors advise otherwise—that the work is original with the authors, is contributed for sole publication in the *Journal*, and is not concurrently under consideration or scheduled for publication elsewhere with substantially the same form and content. Pursuant to U.S. copyright law, authors must sign a copyright release before editorial processing begins. Authors who include data, figures, photographs, examples, exercises, or long quotations from other sources must, before publication, secure appropriate permissions from the copyright holders and provide the editor with copies. The *Journal*'s copyright policy and copyright release form appear in Vol. 18 (1997) No. 1, pp. 1–14 and at `ftp://cs.beloit.edu/ math-cs/Faculty/Paul Campbell/Public/UMAP` .

Language

The language of publication is English (but the editor will help find translators for particularly meritorious manuscripts in other languages). The majority of readers are native speakers of English, but authors are asked to keep in mind that readers vary in their familiarity with vocabulary, idiomatic expressions, and slang. Authors should use consistently either British or American spelling.

Format

Even short articles should be sectioned with carefully chosen (unnumbered) titles. An article should begin by saying clearly what it is about and what it will presume of the reader's background. Relevant bibliography should appear in a section entitled *References* and may include annotations, as well as sources not cited. Authors are asked to include short biographical sketches and photos in a section entitled *About the Author(s)*.

Style Manual

On questions of style, please consult current *Journal* issues and *The Chicago Manual of Style*, 13th or 14th ed. (Chicago, IL: University of Chicago Press, 1982, 1993).

Citations

The *Journal* uses the author-date system. References cited in the text should include between square brackets the last names of the authors and the year of publication, with no intervening punctuation (e.g., [Kolmes and Mitchell 1990]). For three or more authors, use [Kolmes et al. 1990]. Papers by the same authors in the same year may be distinguished by a lowercase letter after the year (e.g., [Fjelstad 1990a]). A specific page, section, equation, or other division of the cited work may follow the date, preceded by a comma (e.g., [Kolmes and Mitchell 1990, 56]). Omit "p." and "pp." with page numbers. Multiple citations may appear in the same brackets, alphabetically, separated by semicolons (e.g., [Ng 1990; Standler 1990]). If the citation is part of the text, then the author's name does not appear in brackets (e.g., "...Campbell [1989] argued ... ").

References

Book entries should follow the format (note placement of year and use of periods):

Moore, David S., and George P. McCabe. 1989. *Introduction to the Practice of Statistics.* New York, NY: W.H. Freeman.

For articles, use the form (again, most delimiters are periods):

Nievergelt, Yves. 1988. Graphic differentiation clarifies health care pricing. UMAP Modules in Undergraduate Mathematics and Its Applications: Module 678. *The UMAP Journal* 9 (1): 51–86. Reprinted in *UMAP Modules: Tools for Teaching 1988*, edited by Paul J. Campbell, 1–36. Arlington, MA: COMAP, 1989.

What to Submit

Number all pages, put figures on separate sheets (in two forms, with and without lettering), and number figures and tables in separate series. Send three paper copies of the entire manuscript, plus the copyright release form, and—by email attachment or on diskette—formatted and unformatted ("text" or ASCII) files of the text and a separate file of each figure. Please advise the computer platform and names and versions of programs used. The *Journal* is typeset in LaTeX using EPS or PICT files of figures.

Refereeing

All suitable manuscripts are refereed *double-blind*, usually by at least two referees.

Courtesy Copies

Reprints are not available. Authors of an article each receive two copies of the issue; the author of a review receives one copy; authors of a UMAP Module or an ILAP Module each receive two copies of the issue plus a copy of the *Tools for Teaching* volume. Authors may reproduce their work for their own purposes, including classroom teaching and internal distribution within their institutions, provided copies are not sold.

UMAP Modules and ILAP Modules

A UMAP Module is a teaching/learning module, with precise statements of the target audience, the mathematical prerequisites, and the time frame for completion, and with exercises and (often) a sample exam (with solutions). An ILAP (Interdisciplinary Lively Application Project) is a student group project, jointly authored by faculty from mathematics and a partner department. Some UMAP and ILAP Modules appear in the *Journal*, others in the annual *Tools for Teaching* volume. Authors considering whether to develop a topic as an article or as a UMAP or an ILAP Module should consult the editor.

Where to Submit

Reviews, On Jargon columns, and ILAPs should go to the respective associate editors, whose addresses appear on the *Journal* masthead. Send all other manuscripts to

Paul J. Campbell, Editor
The UMAP Journal
Campus Box 194
Beloit College
700 College St.
Beloit, WI 53511–5595
USA
voice: (608) 363–2007   fax: (608) 363–2718   email: campbell@beloit.edu